AF616243

Essays on the Foundations of Game Theory

Essays on the Foundations of Game Theory

Ken Binmore

Basil Blackwell

First published 1990

Reprinted 1991

Basil Blackwell Inc.
3 Cambridge Center
Cambridge, Massachusetts 02142, USA

Basil Blackwell Ltd
108 Cowley Road, Oxford OX4 1JF, UK

Library of Congress Cataloging in Publication Data

Binmore, Ken.
Essays on the foundations of game theory/Ken Binmore.
p. cm.
Contents: The aims and scope of game theory—Nash equilibrium—Information—Common knowledge and game theory—Modeling rational players.
ISBN 0-631-16866-4
1. Game theory. I. Title.
QA269.B46 1989
519.3—dc20 89-17507
CIP

British Library Cataloguing in Publication Data

A CIP catalogue record for this book is available from the British Library.

Typeset in 10 on 12pt Times
by P & R Typesetters Ltd, Salisbury, Wiltshire, UK
Printed in Great Britain

Contents

Preface

Recent advances in game theory have ensured its recognition as a key subject across a wide range of disciplines in the social sciences. But, as is often the case when advances come quickly, scant attention has been paid to the foundations on which the theory is being erected. This book does not seek to remedy the deficiency. It is simply a collection of largely free-standing essays, written for the layman, in which some of the relevant issues are chewed over. Most of the material in the earlier essays is straightforwardly instructional at a very elementary level. Later essays become more critical and speculative. The ideal reader will be someone who knows only a little about game theory and thinks that the subject could well be an important one, but is unwilling to put aside his skepticism about foundational issues while spending a long apprenticeship in learning the details of the formal theory.

The essays were written at various times over the last ten years, and in various places including the London School of Economics, the University of Pennsylvania, and the University of Michigan. The first essay, "Aims and scope of game theory," was written at a time when my enthusiasm was at a peak. It was to be the introductory chapter to an ambitious book on game theory which I eventually abandoned half-written when I realized that, in seeking to fulfill the aspirations expressed in the introduction, I was finding it necessary to paper over cracks of increasingly alarming size. Without getting into technical details, the essay seeks to explain what game theory is, what one may hope to learn from studying game theory, and, most importantly, what it is *not* reasonable to expect from its study.

The second and third essays were to be the first and second chapters of another unwritten book. They can therefore usefully be read together as a unit. The idea of a Nash equilibrium is introduced and discussed using simple two-person zero-sum examples. No theorems are proved, but enough technical detail is provided to allow the reader to be able to find the equilibria of simple games. The third essay deals with Harsanyi's theory of games of incomplete information. I hope that this will serve to make it clear that the basic ideas are not at all difficult and hence to make his theory more widely accessible.

The fourth essay, "Common knowledge and game theory." is a joint piece with Adam Brandenburger. It is actually the most recent of all the essays reflecting the fact that it is only very recently that serious thought has gone into the issue of what precisely the players in a game are supposed to know about the circumstances under which they are playing. In particular, what does each know about what the other knows? As with the third essay, I hope that what is written here will serve to demystify a subject that has already become overburdened with unnecessary formalism and misleading jargon.

The fifth and sixth essays were published as a two-part paper in the journal *Economics and Philosophy* but I believe they are worth reprinting here. They argue that what is missing from game theory is any serious attempt to *model the players* and that it is this lack which is largely responsible for the difficulties that have arisen in the foundations of the subject. The sixth essay, "Modeling rational players: part II," makes some tentative proposals on how this gap might be filled, but it will be evident that the stumbling block is not one that will be easily negotiated.

Since the essays were written for a variety of purposes over a period of years, the reader will find the same issues addressed over and over again, sometimes in almost the same words, but usually in greater depth and, I hope, with more sophistication in the later essays. However, definitive answers to the really fundamental questions remain elusive. My own view is that this is largely because we do not yet know the right questions to ask. If this book leads to some questioning of the way the current questions are being asked, it will therefore have served a useful purpose.

But let me end this preface on an upbeat note. My skepticism about the firmness of the foundations of game theory may lead some readers to deduce that game theory can safely be neglected until its practitioners get their house in order. I believe this would be a very wrong conclusion, for the same reasons that it would have been a mistake to neglect mathematical analysis in the eighteenth century. It is true that much remains mysterious or controversial. However, there are many special areas of application in which game theory seems to work very well, although it is not always clear why this should be so. Even in those areas of application in which the theory has clearly been pushed beyond its current limitations, it still cannot sensibly be ignored. Alternative theories of human interaction are typically too protean for it to be possible even to discuss whether their foundational assumptions make them suitable for a particular application or not. In brief, I believe that there is "only one game in town," and that we had better learn how to play it or shut up shop altogether.

Finally, I wish to thank Cambridge University Press for permission to reprint Ken Binmore, "Modeling rational players," *Economics and Philosophy* 3 & 4, 1987 & 1988, which now constitute chapters 5 and 6; and Clarendon Press for permission to reprint part of Ken Binmore, "Why game theory 'doesn't work'," *Analysing Conflict and its Resolution*, edited by P. G. Bennett (1987) in chapter 1.

1 Aims and scope of game theory

> ... draughts and politics, chess and political economy, cards and conversation on nautical matters ... make it the most perfect *concordia discors*.
>
> Charles Lamb, *The Wedding*

1.1 What is a game?

A game is being played by a group of individuals whenever the fate of an individual in the group depends not only on his own actions but also on the actions of the rest of the individuals in the group. Chess is the archetypal example. Whether White wins, loses, or draws depends not only on the moves made by White but also on the moves made by Black. Bridge, poker, backgammon, and Monopoly are further examples. So are baseball, tennis, cricket, soccer, and pool.

For all these activities, the use of the word "game" is familiar. But for the really interesting games it would not be usual to use the word "game" in ordinary English. Consider for example war, the arms race, competition for survival among animals or for status among humans, international treaty negotiations, wage bargaining, elections, or the operation of market economies. In all these activities the welfare of the individuals involved depends as much on the actions taken by others as by themselves, and the activities therefore all fall within our definition of a "game." This usage is not meant to imply that wars are fun or that the arms race is entertaining. It simply reflects the discovery described by Von Neumann and Morgenstern (1944) in their monumental book *The Theory of Games and Economic Behavior* that both parlor games and real-life games pose similar problems and that an analysis which works for the former may very well be relevant to the latter.

Of course, parlor games and real-life games are not similar in *every* respect. One important difference is that it usually does not matter very much whether one plays a parlor game well or badly whereas good play in a real-life game may be a matter of life or death. The analogy which Von Neumann and Morgenstern observed between parlor games and real-life games lies solely in their *strategic* aspects. An explanation of what is meant by this may be helpful. A player seeking the optimal course of action for himself in a game would be wise to consider reducing the problem to its essentials by discarding all detail which is not imediately relevant to the question of what is optimal and what is not. Such detail is at best a distraction and at worst may so obscure matters that the player is not able to come to grips with the problem at all. Once all relevant detail has been stripped away, the player will be left with an *abstract* decision problem. Von Neumann and Morgenstern's observation was that the basic structure of such decision problems is the *same* regardless of whether they are derived from parlor games or from real-life games.

Consider, for example, the game of "matching pennies." This is perhaps not the most intellectually stimulating of games but it is played quite frequently by those with time to kill. Two players each simultaneously show a penny. If both display heads or both display tails, the first player wins a penny from the second player. Otherwise the second player wins a penny from the first. Now consider the not-too-unrealistic naval problem in which an admiral can route a convoy to the north or to the south of an island while his opponent has a single submarine which he can use to cover one and only one of these two routes.

If we abstract away the nonstrategic aspects of these two situations, we can summarize the two games with the matrices of figure 1.1. In both games, each player has two possible *strategies* represented respectively by the rows and columns of the appropriate matrix. The entries of the matrix (called *payoffs*) represent the preferences of the row player. For this example, no significance need be attached to the *numerical* values of the payoffs. Thus the payoffs in the naval game simply mean that the submarine commander prefers that his submarine lie on the route chosen by the convey rather than not but is indifferent to all other issues. Similarly the payoffs in the "matching pennies" game mean that the first player prefers winning a penny to losing a penny and regards other issues as irrelevant. For simplicity we shall assume that the column player has opposing preferences to the row player. (Since the payoffs of the column player may then be taken to be minus the payoffs of the row player, such a game is called *zero-sum*.)

The point of this discussion is that the two games have the *same* payoff matrices. That is, when viewed as abstract decision problems, they pose

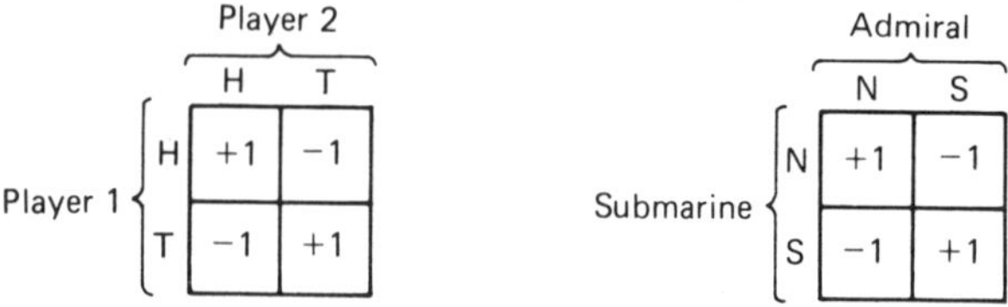

Figure 1.1

precisely the *same* difficulties and a solution to one is therefore simultaneously a solution to the other.

As it happens, the decision problems they pose are very simple and it is not necessary to be familiar with Von Neumann's theory of zero-sum games to see what is optimal for each player given that he acts on the assumption that his opponent will play optimally also.

The argument is based on the assumption that, if optimal reasoning leads a player to a certain conclusion, then this conclusion cannot be secret from his opponent. The reason is that nothing prevents the opponent from mentally placing himself in the shoes of the first player, in which case optimal reasoning will inevitably lead him to precisely the same conclusions as those the first player reached by reasoning optimally. Taking this assumption as given, suppose that optimal reasoning on the part of the first player in the matching pennies game led him to favor the choice of heads over the choice of tails. Then optimal reasoning on the part of the second player would subsume a prediction of this favoritism. But if the second player knows that the first player favors heads, then the second player will favor tails because his payoff is greatest when his choice differs from that of the first player. So far we have assumed for the purposes of argument that optimal reasoning will lead the first player to favor heads and drawn the conclusion that then optimal reasoning will lead the second player to favor tails. But now we can take this conclusion and use it as the first step in a similar argument in which the roles of the two players are reversed. Since the first player's payoff is greatest when his choice is the same as that of the second player, our conclusion this time is that optimal reasoning will lead the first player to favor tails. But this contradicts our original assumption that optimal reasoning will lead the first player to favor heads.

Since the same argument obviously works with the roles of heads and tails exchanged, it follows that optimal reasoning cannot result in a player favoring one strategy at the expense of another. He can avoid such favoritism by delegating his strategy choice to a random mechanism which assigns equal probabilities to heads and tails. For example, he could decide between heads and tails by tossing a fair coin. This is, in fact, Von Neumann's resolution of the game. This is perhaps not a very profound conclusion at which to arrive, especially after so much heavy-handed reasoning, but then the game is a very simple one. Intuitively, each player seeks to prevent the other from predicting his strategy choice and nothing can be harder to predict than a choice made at random.

The games discussed above illustrate a number of points. The first of these is the main point of this section: namely that the abstract decision problems studied in game theory may arise in many different ways, some serious and others less serious. Typically game theorists prefer to describe their results in terms of the less serious applications. Partly this is because less verbiage is usually required, but mostly it is because the less serious applications lack the emotional charge carried by the serious applications and so can be discussed dispassionately. It is however the serious applications, actual or political motives, do no service either to the theory or to society.

The second point is that the question of what constitutes a "solution" of a game is seldom straightforward.[1] Indeed much of game theory is concerned with seeking answers to this question in various contexts. Nor are the solutions of games necessarily easy to understand intuitively once they have been computed. It seems that the human mind is not naturally well equipped for understanding feedback systems. Beginners learning to ski or to helm a dinghy often find that that their natural impulse in an emergency is diametrically opposite to what the situation requires. It is only after falling or capsizing sufficiently often that appropriate reflex actions are learned. Game theory is similar in many ways (except that the conditioning mechanism is seldom so immediate). For example, when Von Neumann's theory of two-player zero-sum games is applied to pot-limit straight poker, the solution[2] requires the opening player always to bet on the first round with J9632 or *worse* but to check (and to fold if bet into) with hands in the range J9632–22J109. Thus optimal play requires always bluffing with a bad enough hand but *never* bluffing with a slightly better hand. Equally surprising is the fact that with an eight-high straight flush (a remarkably rare hand), optimal play requires betting on the first round, making three reraises and then *folding* if the opponent persists.[3] Few regular poker players would find these results easy to predict. They prefer to bluff much less often than the theory recommends and with medium-range hands rather than poor hands. They also prefer to bet very much more boldly with top-range hands. But then, of course, it cannot be the case that many poker players have much experience of poker games in which *all* the players bet optimally, which is the situation to which the Von Neumann theory applies.

The third point to be made is that thinking seriously about game theory requires some experience, knowledge, and sympathy with the manner in which mathematicians reason. This does not mean that a sophisticated knowledge of specific mathematical techniques is required. Some elementary probability, a little linear algebra, and even less calculus will suffice. What is necessary is an understanding of the *nature* of a mathematical argument: in particular, the necessity in a mathematical argument of sometimes taking, as hypotheses, conditional statements which one might very well suspect of being false. Again, this remark should not be interpreted as meaning that a sophisticated knowledge of mathematical logic is required. A reader who is comfortable with the *reductio ad absurdum* argument given above in the analysis of the "matching pennies" game is probably more than adequately prepared.

Unfortunately few social scientists seem to properly understand or appreciate either the aims or the techniques of applied mathematical reasoning. On the one hand, there is a tendency to apply simple models without much consideration to complex situations for which they are ill suited. On the other hand, there is a tendency to denigrate mathematical thinking altogether. Perhaps the next two sections, on the limitations and the aims of game theory, may serve to some extent as signposts indicating a safe course between these two extremes.

1.2 The limitations of game theory

The notion of a game as introduced above is very far-reaching. Almost any situation of interest to social scientists is a game in this sense, and one could therefore take the view that the whole of social science is in principle a branch of game theory. This viewpoint might be defensible on philosophical grounds but has little to recommend it otherwise. It is true that most situations of interest to social scientists can be seen as games but it does not follow that these games can be sufficiently well described for their study to be amenable to game-theoretic analysis, nor that game theory itself is sufficiently well developed to provide an analysis for any game which can be so described. Neither proposition is true and it is important to be realistic about this from the outset since the short history of game theory has already witnessed numerous misunderstandings on this subject. Indeed, immediately after the appearance of Von Neumann and Morgenstern's book in 1944, there was considerable excitement and expectations were high that their theory would have instantaneous and revolutionary applications, particularly in economics. With the advantage of hindsight it is easy to see why these expectations were premature and unrealistic. But the fact remains that there was disappointment, and as a consequence, it is still possible to find elderly economists, ignorant of the more recent advances, who are willing to dismiss game theory as an irrelevance to economics. This is not a very thoughtful point of view but such skepticism is preferable to the gullibility of those at the other extreme who are willing to maintain, for example, that game theory "proves" that certain actions (usually hawkish) are optimal for the arms race. It is almost obvious that such claims are hollow. Just consider the level of understanding of human behavior subsumed in such claims and the implications such understanding would have for human affairs in general. Even a relatively modest claim, such as the existence of a convincing behavioral analysis of the game of Monopoly, would have the most profound consequences for the way our economies are run. This is not to say that game theory has nothing to contribute to the major debates of our time. On the contrary, it is inconceivable that these debates will ever be conducted in a remotely scientific manner without the intervention of game-theoretic ideas. But those who seek to make the theory run before it can walk, either through ignorance or for political motives, do not service either to the theory or to society.

In summary, game theory has great potential. Perhaps it is not too much to say that, unless and until major advances are made in game theory, the social sciences are doomed to remain but a poor relation of the physical sciences. On the other hand, our current state of ignorance on so many relevant subjects imposes heavy limitations on the extent to which game theory as it stands at present can be applied.

To press the latter point home, we now propose to discuss at some length the nature of these limitations. However, in reading what will be a catalogue of ignorance and inadequacy under various headings, it is as well to bear in mind that most theories in social science are so loosely formulated that it would be hard even to know where to begin in enumerating their shortcomings.

1.2.1 Inadequate physical knowledge

The first consideration which limits the scope for game theory applications is the extent of our ignorance about the physical nature of the world about us. We are ignorant across a whole spectrum of relevant issues. We have no detailed understanding of the capabilities and organization of our own bodies. Nor, at the other end of the spectrum, do we have a detailed understanding of the institutions which form the building blocks of our society. Certainly we have much generalized information on these issues. We know fairly well how a muscle works and what nerves are for. Similarly there is no lack of learned books explaining in general terms the way the world banking system works or the relevance of kinship patterns among the tribesmen of New Guinea. But seldom is the information available sufficiently sharp to admit the possibility of a direct application of game theory.

To explain this point it is necessary first to observe that game-theoretic results usually require that we be able to distinguish quite closely between the choices of action which are available to a player and those which are not. It also requires that we be able to describe, again quite closely, how the choices made by the players interact to produce an outcome of the game. In brief, game theory normally requires that a *precise* description of the game is available.

Parlor games such as chess, bridge, poker, and backgammon present no difficulties in so far as precision of their description is concerned. All necessary information concerning the choices of action available to the players and the manner in which these choices interact to produce an outcome is to be found in the official rule-books for these games.

Games of skill like baseball, tennis, soccer, cricket, or pool are quite different. Official rule-books exist for these games but these do not provide all the necessary information. In particular they say nothing about the physical limitations of the players. But a game-theoretic analysis requires this information and is therefore not possible unless a new rule-book can be written which contains not only the official rules but also all relevant information about the physical capabilities of the players. Without such an extended rule-book, one might well find oneself recommending or prediction actions which the players simply cannot carry out. There would be no point, for example, in explaining to a Junior League team that their game would improve if they all swung their baseball bats like Babe Ruth. Nor is the performance of a Fourth Division soccer team likely to be improved by the observation that they would do better if they all played like Pelé. It is necessary to know what is possible for the players and what is not. A generalized assessment of their physical capabilities will not be good enough unless such an assessment makes it clear that one side has a crushing superiority over the other. Interesting matches, i.e. matches for which a game-theoretic analysis would be of interest, will be decided by players operating at the margin of their ability range and presumably it will be a long time before physiology is able to provide the data necessary to determine whether a given standard of performance is or is not within a particular player's ability range. This is not to say that game theory has nothing at all to contribute to the study of games of skill. For example, in principle game theory can be used to analyze the home computer or video versions of such games. In these versions, the real-life participants are simulated by simple computer programs. For games with a considerable strategic element, such analyses might well provide a useful source of new ideas for a team manager. Such games, of course, are just extensions of the table-top games currently played by football managers or potential military commanders in planning strategies. What we have sought to disavow is the notion that game theory can be applied *directly* to real-life games of skill without the availability of adequate physical knowledge of the capabilities of the players.

So far we have considered palor games like chess, bridge, poker, or backgammon for which the official rule-book contains a description of all the rules that an application of game theory requires. We have also considered games of skill like baseball, tennis, soccer, cricket, and pool for which there seems no possibility in the foreseeable future of producing a rule-book adequate for game-theoretic requirements. However, the exciting potential applications of game theory lie in the intermediate area between these two extremes. In this area the official rule-book is not adequate or does not exist at all. On the other hand, there is a realistic prospect of constructing an extended unofficial rule-book which describes the physical and institutional constraints on the players sufficiently precisely to make a game-theoretic analysis feasible.

As an example, consider the game of Monopoly. An official rule-book exists but, aside from the activities closely specified by the official rules, there is also the important possibility that the players may trade their properties on the side. This possibility is admitted by the official rules but the rules do not specify in detail the manner in which this trading is to take place. However, it is not hard to construct some formal rules which adequately describe the manner in which trading is carried out in practice (at least among those who play Monopoly seriously and, as with many diversions intended for amusement only, there are those who take Monopoly very seriously indeed). One specification would require a player to wait until the moment when he is about to rattle the dice-box at the beginning of his turn before he is allowed to propose a trade in properties and/or money from those feasible at that stage in the game. (To enumerate the feasible trades in detail is a straightforward although lengthy business.) The next step would be for the player or players whose participation is necessary for the implementation of the proposed trade to register their assent or dissent. If all agree, the trade takes place and otherwise not. In either case, the proposing player then rolls the dice and the game continues. A more elaborate version would involve counterproposals and counter-counterproposals before the continuation of the game. In this way the unofficial rules governing trade are formalized into official rules in a larger rule-book and a formal game-theoretic analysis becomes possible. Note that the question of the physical capabilities of the players is not an issue in so far as Monopoly is concerned. Nor is it an issue for the class of economic games for which Monopoly will serve as a paradigm. The physical requirements on a player, such as moving counters from one square to another or passing a title deed from hand to hand, are well within the capacity of most of those likely to be found at the Monopoly board.[4]

As remarked above, Monopoly is a paradigm for a large class of real-life economic games in which the official rule-book is inadequate but for which the players' physical capabilities are irrelevant. These economic games do not take place in a vacuum. Institutions have evolved which facilitate and control different types of economic activity. By an institution we not only mean organizations like the Bank of England or the Chicago Wheat Market. We also mean the process of offer and counter offer by means of which a carpet is sold in an oriental bazaar or the arrangements which large companies and trade unions employ when negotiating over wages. In order to use game theory in this context, it is necessary in the first place that the relevant institutions be sufficiently stable and well established that they can be regarded as having fixed rules of operation which the players have no opportunity or desire to violate. The second requirement is that the institution be sufficiently well understood that the relevant rules of operation can be identified and described in a precise manner. In principle, of course, such information can be found by investigation. But this is by no means the same as saying that the information is available in practice. Partly this is because it is not necessarily obvious a priori what snippets of information are likely to be relevant for game theory, and those who are well informed on institutional matters are seldom well informed on game theory and vice versa.

In brief, there is no point in seeking to apply game theory *directly* in economics unless a properly understood institutional framework is present. As noted earlier on games of skill, this does not mean that game theory may not be (and indeed is) *indirectly* useful in the absence of hard, factual information about institutional matters. A game-theoretic study based on hypothetical data about the institutional framework may be useful, for example, as a guide to an empirical worker who is bewildered by the richness of the available data and needs assistance in making judgments about what is likely to be significant and hence worthy of close attention and what is likely to be of only secondary importance.

In economics there are grounds for believing that some at least of the important institutions are sufficiently stable and well established that the writing of an adequate rule-book is not an unreasonable enterprise. An analogous statement can also be made of certain areas in biology. In social and political affairs, however, the situation is more uncertain. With some exceptions, notably elections and the operation of legislative assemblies, institutions seem much more amorphous. The fact that the negotiations over the ending of the Vietnam war ostensibly began with prolonged discussion over the shape of the negotiation table comes to mind. Obviously, the less structure which can be identified in an institution, the less likely it is that game theory can be relevant. Possibly, however, these comments say more about our ignorance of the institutions than about the institutions themselves.

1.2.2 *Inadequate behavioral knowledge*

The writing of a rule-book for a game requires a good knowledge of the physical environment relevant to the game. Such a rule-book will distinguish the actions the players can choose from those which they cannot choose and explain how the players' choices interact to produce an outcome. But the provision of a rule-book is not sufficient in itself. To say something about the play of a game, it is not enough to know what actions are available. It is also necessary to know the method the players use when choosing among the available actions. We shall call such information "behavioral." Thus physical information is required to determine what actions are available while behavioral information determines how a choice is made among these actions. Or, to put the same point slightly differently, physical information determines what the game is while behavioral information determines how it is played.

Ideally game theory would have access to a full-fledged theory of human and animal behavior. This would tell us to begin with what information we need in order to predict behavior and then provide a suitable prediction once the information was supplied. A game theorist would then simply have to determine the parameters of the game with which he was involved, consult the relevant pages in his behavioral theory textbook, and there would be a description of how the game would be played.

Unfortunately no such theory of behavior exists for animals, let alone for humans. However, behavioral psychologists, notably Skinner[5] and his disciples, believe that they have the beginnings of such a theory. They take the view that animals and humans are best seen as stimulus–response machines. The idea is that we cannot know what goes on in the brain of a pigeon or a rat or another human being. We should therefore regard the brain as a "black box" containing an inaccessible mechanism of unknown structure which, when it receives certain stimuli, emits certain stereotyped responses. There is much hostility to this view. Some of this is directed with good reason at the naiveté with which the behavioral psychologists seek to apply what is not an unreasonable basic hypotheses. Certainly there seem no grounds whatsoever for drawing social or political conclusions from the theory as it stands at present. However, probably more hostility is derived from the simple fact that it is not very flattering to have to regard oneself as a "mere" stimulus–response mechanism. Personally I am untroubled by the thought although admittedly I have in mind a self-image somewhat more complicated than a chocolate-dispensing machine.

Whatever a priori opinions one might hold about behavioral psychology, it remains the case that experimental work in laboratories does show that pigeons, rats, and other animals can usefully be described as stimulus–response machines under certain conditions. Nor is there good reason to suppose that humans do not behave very much like laboratory rats when faced with decision problems which they have not thought about very much. For example, if the proverbial man on the Clapham omnibus[6] is allowed to observe the fall of a large number of tosses of a coin which is weighted 70 percent in favor of heads and invited to predict the result of subsequent tosses, then there is evidence[7] to suggest that he will usually guess heads about 70 percent of the time and tails about 30 percent of the time. Precisely the same behavior has been observed in the case of laboratory rats and pigeons suitably rewarded for pressing appropriate levers. To maximise the proportion of successful guesses, one should of course guess heads *every* time.

As a second example, consider the following "take it or leave it" game which has been the object of some recent experimental work.[8] Two subjects who remain separated and totally anonymous throughout and after the game were told that a certain sum of money (sometimes quite a large sum of money) was to be divided betwee them *provided* that both were willing to accept the share assigned to them. The task of assigning shares was then given to one of the subjects and this fact was known to the other subject. If was found that typical behavior on the part of the second subject was to *refuse* his share (in which case both players got nothing) if the share was too small. This, of course, could well be a sound policy if the same kind of situation were liable to arise again and the actions taken by the players during the current game could be monitored by those with whom the players would be matched in the future. It would then be possible to build up a reputation for "toughness" which would be well worth having. But the experimental situation was arranged so that the second player was left with a simple choice between something or nothing, other considerations which might have been relevant under more normal circumstances having been eliminated. Nevertheless, the subjects on the whole took no note of the contrived laboratory conditions but responded machine-like to the stimuli offered them in a manner appropriate, if at all, to an entirely different environment.

To uncover behavior like that described in the examples above requires experimental or empirical studies and, within the context of such studies, the stimulus–response machine paradigm is a natural one. But it is a mistake to be trapped into adopting a laboratory rat model of human behavior. Such a model is only of practical use in those cases where only a small range of controllable stimuli is relevant and the process by means of which these stimuli are converted into responses is reasonably simple. But, although men (and rats too for that matter) may be "nothing more" than stimulus–response machines in some sense, it is obviously true that they cannot always be usefully modeled as *simple* stimulus–response machines.

Consider, for example, the behavior of a Las Vegas casino owner in so far as the running of his blackjack tables is concerned. Blackjack (more commonly known in Europe as pontoon or vingt-et-un) is not a complicated game. The essence of the game is that cards are dealt to the players until they are satisfied, the objecting being to obtain a hand which sums to a total greater than that held by the dealer *except* that hands which sum to more than 21 are "busts," i.e. they automatically lose. (Court cards count as ten and aces as 11.) The casino owner issues instructions to his dealers on the strategy that the "house" is to use and the dealers are not allowed to deviate from this strategy. The point here is that it would be idle to suppose that empirical studies based on the behavior of the gambling population as a whole when dealing blackjack will cast much light on the nature of the house strategy. Doubtless gamblers deal blackjack at home pretty much as they play it in casinos, i.e. with little regard to the state of their bank balances. The same cannot be said of casino owners who would go out of business if they made no profit. It may be that a casino owner is a stimulus–response machine but, if so, he is one for whom relevant stimuli include such factors as the house strategies employed by competing casinos, the memory capacity of inexperienced dealers, and, most importantly, his empirical knowledge of the way the gambling population as a whole play blackjack. Notice that the fact that the final piece of information is part of the data on the basis of which the casino owner makes his decision almost *guarantees* that the play of the gambling population as a whole will *not* be a good predictor of a casino house strategy.

A natural comment is that the house strategies of casinos can be discovered by an empirical study of the house strategies of casinos. Certainly it is not possible to disagree with the virtually tautologous statement that a sound basis for predicting the manner in which player P plays game G is an experimental investigation of the manner in which player P plays game G. But there is a difference between an empirical study and an experimental study. The use of the latter term includes the inference that one can control the relevant variables, while no such inference is implicit in the use of the former term. An empirical study of the house strategies of casinos will reveal only the nature of the house strategies under current circumstances. But for game theory purposes it is usually just as important to know how the house strategies will vary as the circumstances change (i.e. how the response varies with the stimulus). Although it is sometimes possible to devise experiments which will reveal this information, such experiments are typically expensive and sometimes unethical. In any case, it is seldom true that such experimental information is available.

As it happens, few empirical studies can be so easily organized or admit such clear-cut results as a study of Las Vegas house strategies for blackjack. The strategies are very simple with only minor differences, if at all, between casinos. However, Edward Thorpe (1966), among others, has discovered the inadequacy of using such an empirical study as a model for the behavior of the house. Using a computer, he calculated the optimal strategy for a player given the current house strategy. This is considerably more complicated than the house strategy, but not so complicated that it cannot be learned by heart. The use of this optimal strategy leaves the dealer with a very slight edge over the player provided that the cards are dealt at random. But the cards are not dealt entirely at random. In Thorpe's time, four decks were shuffled together and dealt from a "shoe." Consequently less time was wasted on shuffling and more time was left available for taking money from the customers. Thorpe noted, like others before him, that if a count of some sort is kept of what cards have been dealt from the shoe already, then significant statistical information about the cards remaining in the shoe will become available once a sufficient number of cards have been exposed. This statistical information can then be used to shift the bias of the game in favor of the player. However, when Thorpe sought to exploit this advantage, he found that the casinos declined to act like chocolate-dispensing machines and instead reacted vigorously to the stimulus represented by his patronage of their establishments with a variety of measures (some of which make it clear that casino owners interpret the rules of blackjack rather more widely than most of their customers). Finally, after the publication of his book *Beat the*

Dealer, some casinos went so far as the alter the ground rules of their blackjack games.

Notice that a laboratory rat model will not do for a casino owner in this story. Even less will it suffice for Thorpe. Both were engaged in conscious optimizing activity within a conceptual framework of some complexity. To ignore this fact and to insist on a "black box" model for their behavior would be to abandon the structure which renders their behavior explicable on a reasonably simple basis. The vital and fundamental point is that man is a thinking animal and not a laboratory rat or a chocolate-dispensing machine.[9] Sometimes he behaves like a laboratory rat and this behavior needs to be studied. But to claim that the whole truth can usefully be incorporated within this viewpoint is a travesty. Not only can man reason, he can also listen or read and thereby gain access to the thought of others. In particular, he can form theories about how games are played or should be played. Alternatively he can read about such theories in books or newspapers. He may be taught such theories in school. He may even simply note that others are more successful using different strategies and proceed by imitation. Even where behavior seems at its most rat-like (e.g. in the case of stockbrokers or merchant bankers), the players are often acting in accordance with some outmoded theory which, in its time, was new, revolutionary, and quite inaccessible to the mind of a laboratory rat. To repeat a much quoted remark of John Maynard Keynes (1936),

> ... the ideas of economists and political philosophers, both when they are right and when they are wrong, are more powerful than is commonly understood. Indeed the world is ruled by little else. Practical men, who believe themselves to be quite exempt from any intellectual influences, are usually the slaves of some defunct economist. Madmen in authority, who hear voices in the air, are distilling their wisdom from some academic scribbler of a few years back ... soon or late, it is ideas, not vested interests, which are dangerous for good or evil.

The point Keynes is making here is that human behavior cannot be predicted or explained independently of the ideas which humans carry around in their heads. In particular, a theory of human as opposed to animal behavior cannot hope to be widely applicable unless it takes into account the effect which the currently accepted theories of human behavior have upon human behavior. More than this, a theory of human behavior which does not take into account the effect that it would have *itself* upon human behavior if it became widely accepted cannot, of necessity, become widely accepted without contradicting its own predictions. But a theory which could take account of the impact it would have itself upon society would be quite remarkable. In particular it would incorporate a dynamic model of the mechanism by means of which humans form theories in response to external stimuli and how they adapt or abandon these theories with the passage of time.

The remarks above are intended to indicate the intrinsic difficulties involved in constructing an adequate descriptive theory of human behavior. The stimulus–response model is superficially attractive like the glass slipper abandoned by Cinderella at the ball. But the problem is an ugly sister with a foot so immense that it is laughable to suppose that the slipper might fit. The acceptance of this fact of life means that experimental and empirical studies, as presently understood, can only be expected to be useful in situations where humans behave in a knee-jerk fashion without introspection and only minimal calculation. Nor do there seem grounds for anything but pessimism about the prospects for fundamental advances in psychological theory in the foreseeable future. How then does game theory manage given this gloomy assessment of the state of our knowledge on the manner in which humans made decisions?

The answer is that nearly all game theory results so far developed postulate *optimizing* behavior on the part of the players. It is assumed that each player is rational and well informed and makes an optimal choice of strategy on the assumption that all the other players are making their choices on the same basis. Or, to say the same thing a little differently, a game theorist asks: what would each player do if each player were doing as well for himself as he possibly could? As an example, consider the "matching pennies" game of section 1.1. By choosing his strategy on the basis of the toss of a coin, each player can guarantee that he will break even on average. But neither player can improve on this result given the manner in which the other player is choosing his strategy. Thus both players are optimizing given the choice of the other.

Note that this approach short-circuits the behavioral difficulties that we have been belaboring. We no longer ask: how *do* people behave? We now ask: how *should* people behave? To answer the latter question we do not need to know how clever people are, what experience they have had, or how much care and attention they have devoted to their decisions. We only need to be able to describe their preferences over the possible outcomes of the game. Thus a move is good for White in chess independently of whether White is an international grandmaster or a chimpanzee, provided only that both the grandmaster and the chimpanzee prefer winning to losing and regard a draw as intermediate between these possibilities. The fact that the chimpanzee is unlikely to think of the optimal move is irrelevant to the question of whether the move is optimal. As it happens, the immense complexity of chess makes it almost as unlikely that the grandmaster will think of the optimal move except possibly in reasonably simple end-game positions.

Note also that a theory which postulates optimal behavior is one which will not invalidate itself if universally adopted. Given that the other players behave according to the theory, then no particular player will have any motive for deviating from the theory because its recommendation is already optimal for him.

The assumption of optimizing behavior makes it possible to analyze games in a precise no-nonsense fashion. This is very satisfying. It is also satisfying that it leads to a theory which is not self-invalidating. But it is of very great importance not to lose track of the fact that this satisfaction is bought at a price. What is lost is that the players are no longer real people. *Homo sapiens* is replaced by *homo economicus*. Or, if mathematicians will forgive the pun, "real man" is replaced by "rational man". *Homo economicus* (or rational man) is an ideal person invented by economists to model economic agents. Modeling individuals as optimizers certainly generates a theory with considerable potential explanatory power[10] and there is little doubt that situations exist for which the explanation fits the facts. In particular, when handling substantial sums of money on a professional basis, economic agents do seek to optimize given their understanding of the world around them. It is therefore not surprising that economists have found *homo economicus* to be successful as a descriptive model for economic agents in situations which are relatively easy for the agents to understand and to assess. It is more surprising that *homo economicus* is sometimes successful as a model even when the situation is such that the agents clearly do not have any proper understanding of what is going on. (We offer a possible explanation of the latter phenomenon based on evolutionary ideas later on.) However, it is a large step from these successful applications to the assumption sometimes made by macroeconomists that it is realistic to model a large economy "as if" all agents were *homo economicus*, all with perfect foresight,[11] and all optimizing on the assumption that the others are optimizing also.

Before a predictive or even an explanatory role can be claimed for game-theoretic results, it is necessary to have good reasons for why the players can be viewed as *homo economicus*. We need an explanation of how *homo sapiens* became *homo economicus*. Game theory short-circuits the problem of describing human behavior by assuming that players optimize. But game theory offers no explanation of how they became optimizers. The complex and difficult problem of the dynamics of human learning behavior is simply assumed way. But, in seeking to apply the theory, a minimum requirement is an awareness of what has been assumed away and at least the outline of an explanation for why the bald assumption of the theory is reasonable in the context of the proposed application.

There are two types of influence, not necessarily independent, which could well be relevant to an explanation of why *homo economicus* might be a useful approximation to *homo sapiens* in a given context. The first is the influence of *education* and the second is that of *evolution*.

Suppose, for example, that a book were generally available which provided a precise description of optimal play in chess. If a sufficiently important match was to be played, then it would be natural for both players to *educate* themselves by reading the book. Both players would then play optimally and the book would then contain an accurate prediction of their play. Of course, under these circumstances, chess would be a very dull game. Indeed, at least one of the players would probably prefer not to play since he would know from the beginning that it was pre-ordained that he would not win. Fortunately for chess enthusiasts, the prospects of such a book being written are very slim because of the computational enormity of the problem involved. But this is not true of games in general and business schools which have long included rational decision theory on their syllabuses are now beginning to include some education in simple game theory. However, it remains to be seen what impact this education will have on practical game playing even in the business sphere.

Much more important than education, at least currently, is evolution. The "survival of the fittest" is almost a cliché in so far as the animal world is concerned. However, modern developments with a game theory slant, as popularized by Dawkins (1976) in his book *The Selfish Gene*, have added fresh insight by emphasizing that it is the *genes*, or gene packages, which should be seen as the players in the survival game rather than the animals themselves. Those gene packages which confer survival traits on the animals which carry them, given the current environment, are those which tend to survive because the animals which carry them are more likely to live long enough to reproduce. Some gene packages act like optimizers in that the traits they bestow on the animals which carry them are optimal in so far as the survival of the gene package is concerned. That is, if a conscious choice were to be made among the behavioral traits which the gene package could confer upon its carrier, then a choice which optimized the prospects of survival of the gene package would choose the gene package which we have described as an optimizer. Optimizing gene packages will tend to survive at the expense of nonoptimizing gene packages competing for the same role. Of course, what is optimal will change over time. It will be particularly sensitive to the current mix of gene packages in the animal population and this will not be constant. However, in the absence of significant mutations, one might reasonably expect convergence to an equilibrium situation in which all gene packages act as though they were optimizers. In such a situation game-theoretic results will be relevant.

Biological evolution is a familiar notion. Social evolution is a less familiar notion but operates equally remorselessly. In biological evolution, it is important to avoid the mistake of thinking in terms of selection operating directly on the animal population instead of on the gene population. It is, after all, the genes which are the basic unit of replication. Similarly, in social evolution, it is a mistake to think in terms of selection operating directly on the human population instead of on the population of *ideas* which humans carry around in their heads. In social evolution, it is ideas which are the basic unit of replication but, instead of being replicated from one body to another by biological reproduction like genes, they are replicated from one head to another by the social processes of communication and imitation. Here an "idea" might be a complex theory, or it might be a rule of thumb for achieving some purpose, or it might be little more than a learned reflex action. Dawkins (1976) uses the term *meme* to cover all these notions.

The analogy between biological and social evolution is useful but can be overworked. In particular, the processes through which selection can operate are far more varied in the meme population than in the gene population. It remains possible as in biological evolution for meme carriers (i.e. humans) to be eliminated from society and hence for the opportunity for them to pass on their memes to be lost. In the commercial world, those who go out of business usually cease to influence those who remain in business. In polite society, those who behave unfashionably tend not to be invited any more to smart gatherings. In social science, usually only death eliminates the carriers of failed theories. If memes were transmitted from head to head simply by imitation, these considerations would be enough to fuel the evolutionary process. But in human societies certain meme packages, in particular the rational or scientific idea packages, owe their success to the fact that they keep their carriers where it matters by providing them with a more powerful method of acquiring useful memes than simple imitation. This naturally accelerates the evolutionary process very considerably. But it does not necessarily mean that an equilibrium is reached sooner, if at all. This is because the same meme packages which regulate meme acquisition also tend to generate new memes which constantly disrupt the evolutionary flow. The rate of acquisition of new scientific knowledge is commonly said to be exponential. Perhaps this is so. Certainly the generation of new memes seems to be very fast compared with the generation of new gene packages in biological evolution via mutation and sexual reproduction.

All this has implications for game theory. Where meme survival is linked with the welfare of the humans carrying the meme, and when the rate of production of new memes is relatively low, one might reasonably expect convergence over time to an equilibrium situation which the surviving memes can be seen as optimizers. The humans involved in the process need not be "conscious" of what is going on in that the "explanations" offered by the memes on which they base their behavior need not have much basis in physical reality. An individual who practices thrift and hard work may maximize his income as a result and hence become a respected pillar of the community with a consequent boost for the memes he carries. He, on the other hand, may well practise thrift and hard work on the grounds that to do otherwise would be sinful and regard the acquisition of respect and wealth as incidental.

However, one would expect convergence to be more rapid where individuals are "conscious" optimizers so that education and evolution can act hand in hand. Suppose, for example, that, in the blackjack story told earlier, *all* the participants, including the gambling population as a whole and the casino owners, had motivations and capabilities comparable with those of Thorpe. Then the gambling population as a whole would respond optimally to a choice of dealing strategy by the casinos which would force the casinos to respond by altering their dealing strategy to cope with the new situation. This in turn would provoke a new response from the gambling public and so on. The same type of process is familiar to economists under the name of *tâtonnement* in the study of market behavior. Sometimes the term is used in a narrow specialist sense, but the basic idea is that agents adjust their bids and offers in response to the bids and offers of the other agents until equilibrium is reached. It is to this mechanism that economists very plausibly attribute the tendency of markets to balance supply and demand.

The history of chess and bridge provide further examples. Books and magazines describe current practice in the game and, in particular, they report on the play of important matches with analytical commentaries on new developments and proposals for alternatives. The chess and bridge literature is widely read and has a marked influence on the way that games are played. Those who take chess and bridge seriously therefore have little choice but to keep abreast of what is going on and to continually modify the way they play in response. Here we see in perhaps its purest form the manner in which education and evolution can act together to drive a society in the rational direction.

This last point illustrates the long-term prospects for the practical application of that part of game theory (namely most of it) which is concerned with what players *ought* to do in their own enlightened self-interest rather than what they actually *will* do. If such a theory is ever to do more than grace the dusty shelves of academic libraries, it will be because a combination of education and evolution drives society in the direction of the theory. In this process, the existence of the theory itself will be an important factor. A widely applicable theory of games would, of necessity, involve a strong element of self-prophecy in the sense that the existence of the theory itself would be partly responsible for bringing about and stabilizing the events which it "predicts." (Such self-fulfilling predictions are, of course, nothing remarkable on the social scene. The advertising "hype" is a particularly blatant example but similar phenomena are only too familiar in economics.) A game theorist with hopes for the widespread practical application of his subject therefore needs to be something of an optimist (as well as an optimizer) and to hold fast to a faith that the future will see a convergence between theory and practice. He also needs to be something of a propagandist as well as a theoretician since a theory of which the world is ignorant is hardly likely to influence its evolution.

A final world on this last point is necessary and this will return us to our basic theme of the current inadequacy of our knowledge of human behavior. To be a propagandist may be necessary but a line of propaganda which claims too much for a theory is likely to be counterproductive in the long run. In particular, we have outlined above a story which is capable of explaining how it can come about that *homo sapiens* finds himself behaving like *homo economics*. But this story is only a story and its plausibility as a hypothesis depends very much on the circumstances. Certainly its relevance cannot be taken for granted even in biology. To say that evolution *would* lead somewhere under certain conditions is one thing. To say that evolution *has already* taken us there is quite another.

We illustrate this point by considering the game of poker. This game is nowhere near as complicated as chess and is capable of game-theoretic analysis. Have the twin influences of evolution and education brought about a convergence between theory and practice? Briefly, the answer is no. And it is not hard to see reasons why not. In the first place, anyone who has ever played poker will be aware that, although it is necessary to play poker for sums of money which it will hurt to lose, the winning or losing of money is not the only or even the primary motivation of many of those who play. There is the intrinsic excitement of gambling and this is overlaid with a considerable amount of "macho" posturing, i.e. the players seek social payoffs from the game as well as financial payoffs. In addition, there is a constant flow of inexperienced players into the game and an experienced player interested principally in money will obviously prefer to play with these inexperienced players rather than with his peers. But a good player will not play the same against poor players as he would against other good players. In particular, he will *not* play the game-theoretic optimal strategy which assumes optimal play by the opponents because this will not allow him to exploit the bad play of the opposition.

For these reasons, a naive attempt to use a simple game-theoretic model as a predictor in poker is doomed to failure. Indeed, as we saw in section 1.1, optimal play in poker strikes most players as counterintuitive in its recommendations. It is instructive, however, to consider such events as the World Poker Championship where the arguments given against a direct application of game theory are much less compelling. Here play is much closer to that recommended by game theory, particularly with regard to the level and manner in which the players bluff.

1.2.3 Inadequate theory

Most of game theory is concerned with what players ought to do in their own enlightened self-interest rather than with what they will do, and we have discussed at length the limitations that this involves in so far as immediate application of the theory is concerned. But further limitations exist. We do not even yet have a theory of *optimizing* behavior which is adequate to cope with all games, and large areas of game theory remain *terra incognita* desperately in need of exploration.

Some of the difficulties are technical, but the chief problems are philosophical in that it is not so much that we cannot answer the relevant questions as that we often do know the right questions to ask. As an example, consider the glib assertion that game theorists study what happens when each player is rational and well informed and makes an optimal choice of strategy on the assumption that the other players are choosing on the same basis. But what does "rational" mean? What constitutes being "well informed"? Even more troublesome is the circularity built into the assertion. To decide what is optimal for a player, we need to know what is optimal for the other players. But they in turn need to know what is optimal for the first player before they can know what is optimal for themselves. This circularity means that there is no guarantee that the assertion is meaningful. And, even when a satisfactory interpretation can be assigned to the assertion, no guarantee exists that this interpretation is unique.

The difficulties raised by this and other problems are often quite vexing. But they need to be grappled with directly if progress is to be made. Game theory, like all theories, oversimplifies and sidesteps many questions, but there is no room for oversimplification or sidestepping on these theoretical issues. Indeed, the reason for oversimplification and sidestepping elsewhere is to clear the decks so that a direct assault on these questions is feasible.

1.2.4 Inadequate computational ability

It is known that chess has a solution. That is, either White has a strategy which guarantees victory or Black has such a strategy or else both White and Black simultaneously have strategies which guarantee a draw. But to know what a solution *exists* is not at all the same thing as knowing what the solution *is*. One way of finding the solution would be to make a list of all White's strategies and all Black's strategies and then systematically to try each White strategy against each Black strategy, i.e. to play all possible games of chess. Since chess is a finite game, there are only a finite number of possible ways it can be played. Eventually therefore this method will uncover the solution of the game. If this guarantees victory for White, the solution will consist of a strategy for White which wins against *all* Black strategies. Unfortunately for this methodology, even if the process were computerized on the fastest conceivable computer, the universe as we know it would be unlikely to be still in existence before the end of the calculation. Of course, there are more efficient ways of proceeding than simply by playing all possible games. But, even when one considers efficient algorithms, estimates for the time required to compute the solutions of decision problems which are quite simple to describe are capable of making the largest numbers with which cosmologists like to juggle look miniscule by comparison. It follows that there are games whose solution will never be known. Chess is perhaps one of these. This is a fact of life and with facts of life one has to learn to live.

The fact that an optimal analysis of certain games is beyond the capacity of any conceivable computer raises another issue. If no computer can analyze the game adequately, then neither can any player. Should our model of a player therefore take account not only of his physical limitations but also of his mental limitations: in particular, of the data-processing capacity of his brain? Simon is well known in economics for discussing such issues under the heading of "bounded rationality." However, while Simon's observations are relevant they are not in a form which makes them applicable in game theory. Just as an adequate physical description of a baseball player for game theory purposes would require a knowledge of the human body complete enough to make it possible to construct a robot whose physical capabilities were indistinguishable from those of the player it was built to mimic, so an adequate mental description would require a knowledge of the human mind complete enough to make it possible to construct a computer whose mental capabilities[12] were indistinguishable from those of the player it was built to mimic. An optimizing analysis would then seek optimal programs for the computers built to mimic the data-processing capabilities of the players in question.

Such an approach would have many advantages. Computational complexities would be greatly diminished, but the chief advantage would be the potential for coping with the difficulties raised under the heading of inadequate behavioral knowledge. In particular, an optimal analysis for two chimpanzees playing chess would cease to be the same as that for two international grandmasters. However, while the "limited rationality" approach is obviously a direction in which game theory is bound to expand in the near future, only isolated results exist at present. Not only do we not have adequate models of the human or animal mind (although, perhaps paradoxically, these seem closer than adequate models of the body), we do not yet have adequately developed mathematical simplifications for computers (for example, as finite automata or Turing machines).

1.3 The aims of game theory

The wildly ambitious ultimate aim of game theory is to provide a comprehensive analysis of conflict and cooperation in human and animal societies. As explained in section 1.2, the gap between aspiration and achievement is enormous. But, since success in achieving this ultimate aim would simultaneously resolve most of the pressing problems of social science, the existence of the gap is not too surprising. Nevertheless it remains a basic motivation of game theorists to lay the foundation stones for a theory which goes some way in the direction of achieving this ultimate aim.

However, as things stand, we have a theory whose limitations have been stressed at considerable length in section 1.2. So what is this current theory good for? And what grounds are there for supposing that it might provide a basis on which a more comprehensive theory could be built? Obviously, these questions cannot be answered without a knowledge of the theory. What can be done is to give an indication of the *type* of answers which are possible together with some specific examples.

1.3.1 Prediction

Here we intend *precise* hard-nosed predictions as in the physical sciences rather than the soft-nosed predictions typical of the social sciences. By a soft-nosed prediction, we mean a prediction which is consistent with *any* future in the sense that, if it fails to be correct, then the predictor will feel no urge to abandon the theory on which the prediction is based. Instead he will offer explanations of why the prediction failed to be accurate. A hard-nosed prediction, however, is one which, if falsified, *refutes* the theory on which it is based.

As we have seen, the limitations of game theory as it stands at present are such that the hard-nosed predictions it is capable of making will typically be falsified, thus refuting the hypothesis that the players are "rational." It remains an aim of the theory, however, to make hard-nosed predictions where there is reason to suppose that education and/or evolution may have "rationalized" the participants. As an example where education is relevant, one would expect the literature on pursuit and evasion games to have a predictive role in so far as the programming of enemy missiles is concerned. As an example where evolution is relevant, Maynard Smith considers the game in which male dung flies wait at fresh cow pats seeking an opportunity to mate with female dung flies who are attracted to such cow pats as suitable egg-laying sites. How long will a male dung fly lurk at a particular cow pat before seeking a fresher cow pat where females are more frequent? Notice that this is a game-theoretic question because, if all other males wait only a short time before looking elsewhere, a particular male will benefit by remaining at a stale cow pat since, although few females will arrive, his changes of mating with them will be increased by the lack of competition from other males. Here game-theoretic predictions fit strikingly with the experimental data.

A number of misconceptions are common in social science about mathematical reasoning. In particular, "naive positivists" are unable to understand that a mathematical model can have a useful role except in a crudely mechanical and immediately predictive role. Those naive positivists who are not hostile to mathematical models think of themselves as "scientific," a viewpoint they sustain by confusing soft-nosed predictions with hard-nosed predictions. The existence of this school of thought is mentioned only to indicate that there are those who would deny that the game-theoretic aims described next are worthy of attention. However, it is while in pursuit of these latter aims that all the really *interesting* results of game theory have been discovered.

1.3.2 Explanation

In scientific work, predictions are made and tested experimentally to check the plausibility of possible explanations of physical phenomena. But the urge to explain the world around us does not evaporate when the phenomena which puzzle us are not amenable to direct experimentation. Nor is there good reason to suppose that explanations advanced in these circumstances are necessarily without value. Archaeology, cosmology, evolution, and meteorology are all fields where quite the contrary is the case. Of course, explanations which cannot be tested directly are by no means as satisfactory as those which can and they therefore deserve to be treated with some reserve (especially in the social sciences where many popular "explanations" are downright silly). On the other hand, it is as well to bear in mind that we organize most of our daily lives on the basis of such explanations.

The type of explanation which is of particular interest is that which demonstrates that complex patterns of behavior or sequences of events *can* be explained in terms of simpler or more easily understood phenomena and hence that there is no necessity to postulate some "hidden hand" or *deus ex machina* with which to mystify the unwary.[13] Perhaps the best contemporary example is the explanation that molecular biologists now offer for the origins of life on this planet through the creation of self-replicating molecules in a primeval soup. This they envisage as a random but highly probable event (see Monod's (1972) *Chance and Necessity*). Enough molecular biology is now known to make this at least a plausible explanation but there is no way to test the explanation directly. Nevertheless, given the role that religion has played and continues to play in human affairs, there is little doubt that the existence of the explanation is of immense importance and interest. A more traditional example, of course, is the replacement of the Ptolemaic epicycle explanation of planetary motion (based on the supposed perfection of the circle) by Keplerian ellipses and the Newtonian theory of gravitation.

Of course, the fact that a phenomenon *can* be explained in a simple way does not mean that this explanation is correct. It is easy to think of phenomena which admit a simple and straightforward explanation but for which the actual explanation is very complicated, as those husbands who have been discovered in compromising circumstances by their wives known only too well. Thus, although molecular biologists offer a simple explanation of the origins of life, it remains possible that life was created by God just as described in the Bible or that the Earth was seeded from outer space as some physicists propose. Occam's razor is often quoted in this context. It asserts that, where rival explanations are available, that which assumes least should be preferred. This is sometimes interpreted to mean that complicated explanations should be rejected in favour of simple explanations. A more reasonable interpretation is that simple explanations should not be rejected in favor of complicated explanations without additional evidence. A rider is that, when considering possible explanations for something puzzling, one should actively seek for simpler explanations before committing oneself to whatever explanation is currently popular. This is particularly important when the popular explanation attributes the phenomenon partly or wholly to the action of some black box mechanism, i.e. an entity or principle whose reason for existence and manner of operation is left unexplained. A tapping at the window in the deep reaches of the night is easily attributed to the malice of the walking undead but this explanation is best evaluated in the light of simpler possible explanations.

To what sort of phenomena is it possible that game theory might provide "simpler" explanations than those currently popular? We shall propose two classes of appropriate problems, both very important. The first is essentially sociological in character and the second economic.

Much of human behavior in society is attributed to the influence of morality. Those who live in New York may not agree, but to a considerable extent people tell the truth, keep their word, respect property rights, and help old ladies across the road. Why do they behave this way? One answer is that otherwise they suffer from feelings of guilt, with the source of the guilt variously attributed to the existence of a built-in conscience or to childhood conditioning. A second and not unrelated answer is that it is in an individual's self-interest to protect his reputation. But these explanations involve black box ingredients. This is most obvious in the case of the postulated existence of an entity called the "conscience" (and even more so if one examines such Freudian concepts as the id or superego). The explanation in terms of childhood conditioning is less of a *deux ex machina* but still leaves unanswered the question of *why* parents should condition children beyond what is dictated by the selfish interests of the parents. Even the explanation in terms of "reputation" begs the issue of *why*, for example, a reputation for generosity and open-handedness should have come to be valued by society. Certainly there is no necessity that such a reputation should be valued, as is demonstrated by the much quoted example of the Ik of Uganda who operate a very much more selfish social norm that is common amongst Western societies.

Doubtless all these explanations and others are relevant in so far as the maintenance of the moral fabric of our society is concerned. But, as we have seen, they are all ramshackle to some extent in that they raise nearly as many questions as they answer. Game theory can help in this context by providing putative answers to some of these subsidiary questions and hence supplying the glue or the nuts and bolts necessary to hold the ramshackle explanations together.

We give two examples very briefly. The first concerns "altruism." This may be hard to explain in human societies without appeal to a "moral" black box but its appearance in animal societies is even more mysterious. However, if we can explain altruism in animal societies without recourse to a black box, do we really need a black box to explain altruism in human societies? Consider the behavior of the social insects which is presumably almost entirely genetically determined. Note in particular the altruistic behavior of the worker honeybee as she toils selflessly in the interests of the hive. Such selflessness extends to committing almost certain suicide by stinging raiders who threaten the hive's honey supply. How is this consistent with rational, optimizing behavior? Here it is necessary to recall two things. The first is that it is the *genes* which the bee carries which are seen as the players; i.e. they are seen as programming the bee's behavior as though seeking to optimize their survival prospects. The second is that worker bees are *sterile*. They cannot therefore propagate their genes directly as with most animals. Instead they are programmed to tend and protect the brood of the hive and therefore to promote the survival of those genes that they share in common with their sexually active brothers and sisters. Not only is it possible to explain the apparently altruistic behavior of the worker bee in terms of "selfish genes," it is also possible to offer a simple explanation of why such behavior evolved. *Hymenoptera* (including ants, bees, and wasps) have curious genetic arrangements as a result of which two sisters have more genes in common than a mother and a child. A worker bee (which is genetically female) therefore promotes the survival of the genes it carries better by tending a sexually active sister than it would do if it were sexually active itself and tended its own child. As Dawkins (1976) comments, it is presumably no accident that true sociality, with worker sterility, seems to have evolved no less than 11 times *independently* in the *Hymenoptera* and only once in the whole of the rest of the animal kingdom, namely in the termites.

The example of the social insects is very striking and it is not hard to draw parallels in human society. Instead, however, we shall give a more mundane example involving the parlor game "Diplomacy" since this is less likely to be controversial. This game involves a map of Europe ca. 1900. Each of up to seven players represents a major country of the period with appropriate numbers of armies and fleets. At each move, all the players write down the actions they propose for their armies and fleets and these instructions are then carried out unless they provide impossible as a result of the instruction of other players. Between moves the players have the opportunity to coordinate their strategies in secret or semi-secret negotiations. However, nothing binds the players to anything they may agree during these negotiations. They are free to break their word and very there are considerable advantages to be gained by so doing. For example, Germany and Austria–Hungary may agree to a joint onslaught on Russia (while doubtless reassuring Russia that they plan nothing more than a mutual defense pact). However, Germany may then treacherously invade the Austrian provinces left defenseless by the Austrian movement into Russia. Beginners occasionally give voice to expressions of moral outrage at such treatment. These are typically greeted with much hilarity on the part of the more experienced since the game has no point if such moral issues are allowed to intrude. Nevertheless, by and large, successful players do keep agreements, tell the truth, and sometimes help out a neighbor even when there is no obvious immediate advantage. Why? The game-theoretic answer is that if players estimate the chances of another player's keeping a proposed agreement partly on the basis of his past record (i.e. his reputation), then it becomes worthwhile to keep agreements in order to establish a suitable reputation. Nobody will cooperate with a player who has a bad reputation for betrayals and an isolated player is soon eliminated. At the same time, this behavior has a self-reinforcing quality in that the fact that players tend to keep agreements in order to acquire a good reputation confirms the original expectation that players with a good reputation keep agreements.

In real life people are seldom conscious of why social evolution has programmed certain behavior in themselves and are quite comfortable with the most facile of explanations. In Diplomacy, a good player cannot afford to be unconscious of his motives for establishing a reputation because victory depends on choosing just the right moment to abandon the cautious reputation-building policy in order to stab an erstwhile ally in the back. For this reason, Diplomacy and other more easily analyzed games (such as the "prisoner's dilemma" game) provide a useful paradigm in seeking simple explanations for the manner in which social role-playing becomes an entrenched feature of our social life.

The problems from economics considered next have a rather different flavor. They concern "efficiency." Here it is important to appreciate from the outset that economists use the word efficiency in a manner which would not occur immediately to the layman. A situation is said to be efficient if any alteration would injure at least one of the economic agents involved. This definition is due to Pareto and the fact that the term is being used with the precise meaning he proposed is usually emphasized by describing the situation as *Pareto efficient*. Sometimes a Pareto-efficient situation is said to be Pareto "optimal," which is highly misleading since it tempts one to suppose that the aim of a socially concerned economist is attained once he has achieved Pareto "optimality." But is it necessarily "optimal" in a social sense to distribute sweets among a group of schooldhildren so that one receives them all and the remainder receive none? Nevertheless, if the schoolchildren are selfish, this distribution is Pareto efficient.

Related to the question of Pareto optimality is the operation of the free market. Few would deny that there is much empirical evidence to suggest that, across a wide range of activities, the free market is at least as good a mechanism for distributing goods and services as the planning activities of the type of bureaucracies with which we are currently saddled. But there are those who go beyond this and maintain, mostly implicitly but sometimes explicitly, that the market mechanism should be regarded as paramount and that interference with its operation is *necessarily* socially undesirable.

This Panglossian view is not without theoretical support in that game-theoretic mathematical models of simplified economies can be constructed in which selfish optimizing behavior by all economic agents results in a Pareto-optimal distribution of labor, goods, and services. The prototype of such models is that of Arrow and Debreu (see Debreu, 1959). An ownership pattern is assumed in which consumers begin with an initial endowment of goods and shares in existing production units. Taking the price system as given, the production units then choose a bundle of outputs (which will include negative quantities such as labor and raw materials) in order to maximize profit. Again taking the price system as given, consumers can now compute how much their initial endowment is worth and how much their share ownership will yield. These together tell a consumer how much he can spend; i.e. they determine his budget constraint. Consumers are now assumed to purchase the bundle of commodities (which may again include negative quantities such as labor and raw materials) which they most prefer given that they cannot violate their budget constraint. The immediate issue is whether all these selfish optimizing decisions, chosen quite independently of the decisions made by other agents, are compatible. Will the producers realize the profits they plan? Will the consumers find the goods they propose to purchase available? More briefly, will supply be adequate to meet demand? Arrow and Debreu show that, under appropriate conditions, the answer will be "yes" provided that a suitable price system prevails. Such a price system is called an equilibrium or *Walrasian* price system. What is more, it is easy to deduce[14] that, at a Walrasian equilibrium, the pattern of distribution and production of goods is Pareto optimal. Arrow and Debreu do not provide a formal explanation of how a Walrasian price system is to be brought about but sometimes economists speak of the "hidden hand" or of an "auctioneer" in this connection. But what they really have in mind is a *tâtonnement* process, as discussed in section 1.2, under which bids and offers are adjusted in response to the bids and offers of others until equilibrium is reached.

The Arrow–Debreu model shows a number of things. Perhaps the most interesting is that prices can be very effective as carriers of information about the economy as a whole. The point we are concerned with here, however, is that it shows that Pareto efficiency *can* be achieved as a result of selfish optimizing behavior without the intervention of a central authority. This is by no means a self-evident result. Indeed, to the layman it seems paradoxical that competitive behavior could result in socially efficient results (no matter how weakly "socially efficient" is interpreted). To some economists, on the other hand, the notion that free markets yield "optimal" outcomes is not so familiar that it is the countersuggestion that rational behavior on the part of independent agents may lead to *inefficient* results which has come to be regarded as paradoxical.

The traditional explanation of how this can come about is given in terms of the prisoner's dilemma game. Similar results can be rephrased in terms of oligopolistic competitition between several firms who are large enough that they do not take prices as given, as in the Arrow–Debreu model, but are aware that their production decisions will significantly affect prices. However, it is more instructive at this stage to consider wage bargaining and, in particular, strikes and lockouts.

It is often said that, in a "rational world," strikes and lockouts would not be observed. The negotiators would take into account the damage that each has the potential to inflict on the other (and themselves) from the outset. An agreement would then be reached immediately without any unfortunate unpleasantness. Also regarded as irrational is the common practice especially on the trade union side, of fixing firm and supposedly immovable dates at relatively infrequent intervals at which to review their position. If the firm makes a new and better offer, there may then pass up to a week before it is even considered. Obviously strikes and lockouts are inefficient and anything that prolongs a strike or a lockout increases its inefficiency. But are these activities necessarily *irrational*?

The answer is "no." Game-theoretic models show that if one party to the negotiation, or both, are ill informed about the extent to which a strike or lockout will injure the other party, then a player who is actually "strong" in that he can endure a long strike or lockout with relatively little injury will find it optimal to signal this fact by initiating a strike or lockout. It is of the essence of the matter that this signal be costly not only to the other party to the negotiation but also to himself: so costly in fact that it would not be worthwhile for a "weaker" player in his role to mimic the signal in the hope of being mistaken for a "strong" player. The "strong" player will then have established his credentials and reap the benefits of being identified as a "strong" player. As far as fixing a firm timetable for reviewing the situation is concerned, game theory models show that this can be advantageous even when both sides are fully informed about the other. When the firm is considering a union offer, the knowledge that a further week at least of strike will follow a refusal increases the pressure on the firm to accept.

None of this is intended to imply that wage bargaining is normally, if ever, carried out rationally: only that certain behavior is capable of rational explanation and hence that it is irrational to dismiss it as irrational without further evidence.

1.3.3 Investigation

A first step in investigating something which engages the attention is to consider possible explanations as described above on the basis of the data to hand. These possible explanations then provide a guide as to what new data are likely to be relevant and to the character of the experiments which might be useful.

But the collection of data and the running of experiments in the social and biological sciences is often a task of very great difficulty. Even where the required data are available in principle or a properly controlled experiment could be run in theory, it may be that high costs or ethical considerations or the hostility of interested parties make it impossible to run the experiment or gather the data. For this and other reasons it is sensible to borrow a research technique from pure mathematics. The same technique has been used by physicists since at least the time of Archimedes.

A mathematician looking into an area for the first time usually has some hazy notions of what type of theorem might be true. Experience shows that these notions tend to remain hazy if the mathematician does no more than contemplate the problem in its full generality from a safe distance. A common research strategy is therefore to look closely at simpler problems in the hope that these will suggest how the hazy notions floating at the back of the mind can be refined into precise conjectures capable of proof or disproof. Sometimes these simpler problems are obtained by particularizing the original problem, i.e. by looking at specific examples; sometimes by generalizing through the suppression of details thought likely to be irrelevant; sometimes by studying a parallel problem from a different mathematical framework with a view to seeking analogies (see Pólya, 1962, or Lakatos, 1976).

Once a precise conjecture has been formulated, there comes the question of proof. Again insight is sought by looking first at simpler problems. Particularly important at this stage is the search for *counterexamples*; i.e. one actively seeks to construct examples in which the conjecture is false. If a single such counterexample is found, the conjecture is refuted and a new conjecture must be formulated which is consistent with the counterexample. If no counterexample can be found, the effort expended in the search is seldom wasted. Much insight into the structure of the problem can be obtained in this way without which the construction of a proof is unlikely.

This discussion is intended to indicate how the study of examples in pure mathematical research, especially the research for counterexamples, mimics the experimental techniques of the natural scientist. Indeed, Einstein used to call such studies "mind-experiments." When such mind-experiments are employed, as with Einstein, outside pure mathematics, it does not greatly matter whether or not the hypotheses on which they are built are realistic or even realizable. The purpose of the mind-experiment is usually to test the *internal logic* of a theory rather than to verify or to challenge its predictions. For example, suppose an economist maintained that, under a *laissez faire* regime, an economy will necessarily stabilize at an equilibrium without involuntary unemployment. Doubtless he will have arguments with which to support this assertion. Since our knowledge of the world is incomplete, these arguments will have to apply not only to the world as it is but also to a variety of other worlds which might have been. In particular, most economists would agree that their arguments would remain valid in a world in which *homo sapiens* were replaced by a *homo economicus*. Such an agreement makes it possible to seek to construct a mathematical model of a theoretical economy in which the hypotheses of the economist are valid but his conclusions are false. The model is then a counterexample to the general theory; i.e. it serves as a falsifying experiment-substitute. It refutes not the predictions of the theory but the validity of the supporting argument.

The fact that a mathematical model with unrealistic hypotheses can be useful is not widely understood and such models are often treated with derision by "naive positivists." Such derision might equally well be directed at Einstein for postulating that Swiss tramcars might carry junior patent clerks to their work at the speed of light.

In order to introduce an apt illustration of the role of mathematical models as logical counterexamples, the following quote from Harsanyi (1977) is offered:

> ... Marxist writers and other social scientists who have stressed the importance of conflicting class or group interests have found little room in their theories for the common interests of different social classes or groups. Conversely, functionalists and others stressing the importance of the common interests in society have found little room for conflicting sectional interests. Indeed a well-known German sociologist, Ralf Dahrendorf, has gone as far as explicitly asserting the strange doctrine that the general interests of society and the conflicting sectional interests of various social groups simply *cannot* be used as explanatory variables within the same model ...
>
> Dahrendorf (1958).

However, as Harsanyi points out, game-theoretical models explicitly take account of any conflicts of interest that may exist among different players and simultaneously of any common interest they may have in reaching a peaceful agreement and in cooperating to achieve their shared common objectives. Virtually any game-theoretic model, no matter how unrealistic its hypotheses, is therefore a logical counterexample to the Dahrendorf "doctrine."[15]

1.3.4 Description

There is a great deal in this world which we do not understand very well but about which we are only too willing to speak at enormous length. Politics provides a good example especially since the words expended on this subject can and often do have an overwhelming influence on the lives of ordinary people. But consider the paucity of the available vocabulary. Convention, for example, requires us to locate politicians along a one-dimensional left–right spectrum which we all know to be totally inadequate as a means of describing the complexity of contemporary political opinion. But, by and large, we submit unthinkingly to this crippling and outdated conceptual straitjacket because no better descriptive system is realistically on offer. Even words which began their lives with useful and reasonably precise meanings are debased into meaninglessness by systematic abuse. Consider, for example, the word "fascist" as used by those who think of themselves as "on the left" or the word "free" by those who consider themselves "on the right."

One of the things which game theory has to offer is an extensive and dispassionate vocabulary within which the issues of conflict and cooperation which arise in a political, sociological, or economic context can be discussed with a minimal risk of misunderstandings arising from individuals' being taken to hold complex but ill-defined packages of political opinions because they express themselves in words which have an emotional charge for those listening. Aumann (1987) has pointed out that the provision of a useful vocabulary for discussing problems of conflict and cooperation (i.e. games) is essentially an act of *classification* and draws attention to the foundations of the modern science of biology in the classification system of Linneaus by means of which animals or plants are described in terms of their *species*, *genus*, *family*, etc. Such classification is sometimes said to be a "science" (in which case the word *taxonomy* is often substituted for the word classification). However, as Aumann observes, what *matters* about a classification system is whether it allows one to make the distinctions or to note the similarities which one feels relevant or, perhaps more important, whether it draws to the attention similarities or distinctions which seem significant but which might have passed unnoticed without the classification system. The sensible question

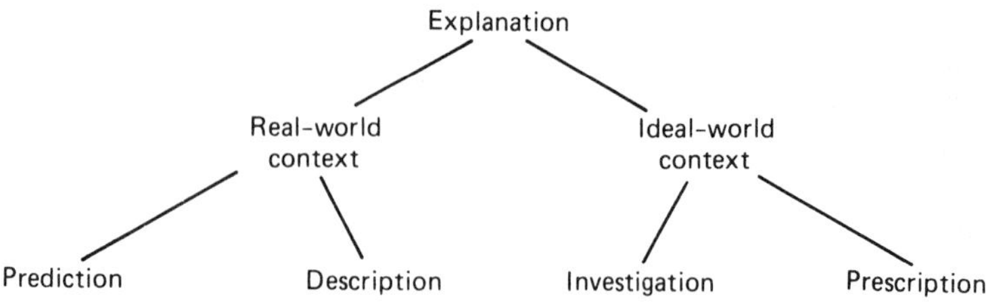

Figure 1.2

to ask of a classification system is therefore not so much whether it can be said to be "true or untrue" but whether it "works." In this it resembles an office filing system or a complex computer program more than a "scientific" theory. For example, one might offer the following hierarchical classification of the aims of game theory as presented in this section (figure 1.2), but to ask whether this classification is "true" is to miss the point. The issue is whether a particular individual finds it a help in arranging his thoughts and this will depend on his background and interests.

Returning to the specific contribution which game theory has to offer in respect of the vocabulary with which we describe the world around us, it is easy to see the usefulness of drawing analogies between real-world situations and particular games or types of game. Military theorists for example have long found it useful to draw strategic parallels with chess, while poker has been a particularly fruitful source of useful words for describing situations which arise during negotiations. Think for example of "bluffing" and raising the stakes." More technical terms have also begun to make an appearance. Sociologists speak of situations not being "zero-sum" when they wish to point out that mutual benefits are available for cooperation. Political scientists use the prisoner's dilemma game as a paradigm for a wide class of situations in which cooperation would be mutually beneficial if the players could rely on each other to honor agreements but in which, whatever worthwhile agreement may be reached, each player has an inducement to cheat regardless of whether the others are honorable or not.

This last example highlights an important distinction that game theory makes concerning the context within which games are played. A preliminary classification distinguishes *cooperative* and *noncooperative* contexts. In a cooperative context, agreements are *binding*, either because of legal sanctions or because the players' reputation for integrity is overwhelmingly more important to them than any advantage that can be gained by cheating on a deal. In a noncooperative context, agreements are not binding or only partially binding. This classification is simple but of considerable importance. It is not obvious, for example, to the man on the Clapham omnibus or indeed to many politicians that there is a fundamental difference between a negotiation amongst businessmen over a contract to supply ball-bearings and a negotiation between diplomats over an international treaty to regulate trade. The latter is a game played in a noncooperative context and both parties are likely to deviate from an agreement as soon as it no longer suits their convenience. The agreements available to the diplomats are therefore more restricted than in the case of the businessmen whose contract will normally have legal backing. The diplomats may only realistically be able to settle on agreements which neither party will afterwards *want* to violate. Such an agreement will usually be better than no agreement at all, but not necessarily (as the prisoner's dilemma game was constructed to illustrate). If this is not understood and expectations are unreasonably high about what is achievable, then even the limited scope that exists for agreement may be put at risk. Of course Neville Chamberlain's famous agreement with Adolf Hitler is still sufficiently well remembered that naiveté over dramatic issues such as disarmament is fairly limited. But, for example, the repeated attempts to renegotiate the terms of British membership of the European Economic Community continued to arouse unreasonably strong feelings on both sides of the English Channel long after Britain joined the Community.

The cooperative–noncooperative distinction is only one of a number of abstract game-theoretic classification concepts which are of immense importance in so far as the description of the world around us is concerned. A less abstract contribution from game theory concerns the description of "power" in legislative assemblies.

Counties in New York State are run by Boards of Supervisors. The supervisors do not have equal votes because the size of their electoral districts are often markedly different. Instead their votes are weighted to reflect the number of citizens they represent. This is similar to the arrangements at shareholders' meetings except that the weights are not directly proportional to the populations of the supervisorial districts. A legislative assembly operating with such weighted votes certainly seems "fair" in the abstract but a closer examination is appropriate.

Nassau County, for example, had six supervisors in 1964 whose votes carried the weights 31, 31, 28, 21, 2, and 2. Note that the vote accorded the fourth, fifth, and sixth supervisors by this scheme is worthless provided there are no abstentions or absences. Under these circumstances a motion will be carried if and only if it has the backing of two of the first three supervisors; i.e. the outcome will be the same as if the weights were 1, 1, 1, 0, 0, 0. Shapley (1953) has coined the apt term *dummy* to describe individuals who are powerless in the sense outlined above for the fourth, fifth, and sixth supervisors.

Presumably the fact that some of the supervisors were dummies was not obvious a priori since otherwise an alternative system would have been adopted. Indeed, in 1971, Nassau Country modified its arrangements and no longer has dummies. (The weights were kept constant but a total vote of 63 became necessary to ensure the passage of a motion instead of the previous 58.) It is equally not a priori obvious to the man on the Clapham omnibus nor to many politicians that, in an ordinary one-man one-vote legislative assembly containing three solidly disciplined parties holding seats in the ratio 5:4:3, each party has essentially the same power as if seats were held in the ratio 1:1:1. But nothing complicated is necessary to make this observation. All that is required is to ask the necessary question. The answer is then immediately obvious. But, if the notion of "voting power" is not part of a person's vocabulary or conceptual framework, then the necessary question is simply not asked.

Shapley and others (see Owen) have sought to close this gap in our vocabulary by introducing formal definitions of "power" based on game-theoretic considerations. Shapley's index, for example, assigns 98 percent of the available power at the United Nations Security Council to the five permanent members. It assigns a Californian voter around three times as much power as a voter from the District of Columbia in a presidential election. (Recall the existence of an "electoral college.") It assigns nearly 50 percent of the available power in a legislative assembly to a large party holding roughly one-third of the seats provided that the other parties are small and sufficiently numerous. The erstwhile dominance of the Israeli Knesset by the Labor Party is usually quoted in this context. In none of these cases is the distribution of power obvious to a casual glance.

The point in this discussion is not so much whether the numbers Shapley assigns are the "correct" numbers. Even the meaning of "correct" is open to debate in this context. The question is whether Shapley's definition helps us to say things which we would like to say but otherwise would not find it easy to say because of their entanglement with a whole mess of other political and often emotional issues.

As a further example, consider the British Trade Union Congress. Union leaders vote by holding up cards on which are inscribed very large numbers (often in the millions) representing the nominal membership of the union. Since the issues voted on are often of very significant social, economic, and political importance, the question of the distribution of power is worthy of attention. Only when it is recognized that the game-theoretic structure of the voting arrangements confers a power on the union leaders which is *not* adequately measured by the nominal size of their membership does it become possible to ask: What would be an "equitable" weight to attach to a union leader's vote? The square root of his membership is often mentioned in this connection with good reason, but this anticipates that we have to say about the prescriptive role of game theory.

Before leaving the *descriptive* role of game theory, we wish briefly to make the final point that even models with self-confessedly unrealistic elements can be important in this context. Harsanyi (1977), for example, points out that any worthwhile account of Napoleon cannot fail to mention that he was a brilliant military strategies but a less than adequate diplomat.[16] But these descriptive remarks implicitly require envisaging what would have happened in an imaginary world in which Napoleon was replaced by a specimen of *homo economicus.* (Doubtless Josephine would have been equally disappointed.)

For an example in which game theory is more directly relevant, consider the descriptive notion of "exploitation" as used by Marxists. Presumably most people would agree that the word is meaningful but only by virtue of seeing the world around us as a distortion of an ideal world with a very different ownership pattern and authority system and for which a game-theoretic analysis would be entirely apt as a result of the rationality of its inhabitants. Rawls (1972) is well known for his theory of social justice which is based on a similar idea. He asks what society would result from a Rousseau-like social contract if the parties to the contract had no a priori information about which role they would occupy within this society. He then offers an analysis of this "original position" and considers, with some provisos about individual freedoms and the like, that the contract would involve maximizing the utility of the least well-off. But this again returns us to the *prescriptive* role of game theory.

1.3.5 Prescription

Just as doctors prescribe courses of treatment with a view of promoting the health of their patients, so social scientists are in the habit of prescribing various courses of action with a view to achieving certain aims within society. The nature of the prescription, of course, depends upon the aim to be achieved. Game theorists in particular are well placed to offer advice on two distinct types of problem. The first is the obvious issue of how best to play a given game. However, the rules of a game cannot always be treated as given and a secondary problem is to determine the nature of the game to be played with a view to ensuring that its play will not lead to socially undesirable outcomes or, at least, to outcomes which conflict with the objectives of those for whom the game is designed.

As far as the first type of problem is concerned, a good example is the book *Beat the Dealer* mentioned earlier in which Thorpe (1966) described how to win against the blackjack dealing strategy then employed by Las Vegas casinos. Investment and financial councillors, political advisers, and economic and management consultants are all engaged in what is an essentially similar activity and are therefore all game theorists at least to some extent. But the type of game theory they study is rather special in that they typically proceed by making a judgment about how the other players in the game (e.g. the casino or the market) will behave and then optimize taking this judgment as given. In essence therefore they model the problem which they are faced as a game with only one player. Such problems are usually studied under the heading of *decision theory* on which there is a large and important literature. *Game theory* on the other hand is usually understood to be concerned with games with at least two players. This distinction is somewhat artificial. Certainly game theory cannot be studied without some decision theory being learned along the way. On the other hand, it does serve to emphasize that game theorists are interested in offering advice when the behavior of other agents *cannot* be taken as given and it must be assumed that they also are in receipt of good advice. This point was made earlier and so brief mention will only be made of the example of the advice to be offered to a large firm competing with a small number of other large firms in a given market, or the advice one might publish concerning strategic voting in an election with three significant parties.

The rather artificial distinction which exists between decision theory and game theory has been mentioned. With regard to the second type of problem about which game theory can offer useful prescriptive advice there is again a somewhat artificial distinction between *social choice theory* and *game theory*. The former is concerned with what are essentially ethical questions of social justice. What should one do, for example, if faced with the choice between producing a large cake which has to be unfairly divided and producing a much smaller cake which could be equally divided? Game theory is closely connected with questions of this sort. Indeed, it did not prove possible to discuss the *descriptive* role of game theory without touching on these questions. This discussion is now continued by outlining a traditional example concerning the "fair" allocation of airport landing fees. The problems which arise in arbitrating over labor disputes are very much more interesting but remain a vexed and controversial issue.

There seems no difficulty about assigning the proportion of a landing fee which goes towards the running costs of an airport and our concern is therefore entirely with the proportion required for the amortization of the original capital investment. The size of the original capital investment is largely a function of the length of the runways and it therefore seems reasonable that aircraft which need a longer runway should pay more. But how much more? One way of proceeding, which is currently increasingly popular with academic accountants, is to use the *Shapley value* which has already been met as an index of voting power.

In the latter context, one may think of testing voting strengths by hypothesizing issues so divorced from current political realities that all possible alignments of the voters on these issues can be taken to be equiprobable. Such an alignment is to be thought of as ordering the voters according to the intensity of their support or opposition to the proposal. In such an ordering a particular voter on the winning side will be critical in that, if all those who voted on the winning side but felt less intensely than he were to change their minds, then the motion would still be carried but, if the critical voter also changed his mind, then the motion would be lost. The Shapley value in this context is then the probability of being critical. In the airport-costing context, the alignment of voters may be replaced by an imagined order of arrival of aircraft. Instead of seeking a critical voter, one asks, for each aircraft as it arrives, what would be the notional cost of increasing the length of the runway to accommodate the latest arrival given that a runway which accommodates all previous arrivals has already been constructed. The Shapley value for a particular aircraft, or type of aircraft, is then found by averaging such notional marginal costs, taking all arrival orderings as equally likely.

Such applications are game-theoretic in so far as, like the Rawlsian theory of justice mentioned earlier, they have in the background an imaginary game used as a standard of reference. However, the question of whether the proposed allocation of costs or distribution of power is "fair" is a matter for the exercise of ethical judgment rather than for game theory.[17] Such judgments are of course of immense interest and importance but their study is usually regarded as belonging to social choice theory rather than to game theory.

A subject which belongs very firmly in game theory, however, is the important question of the *implementation* of ethical judgments. Here the problem is, taking a set of social aims as given, how should society be reorganized to bring these aims about. More precisely, there are many situations in which an individual or an institution controls the rules of games played by other people. The government for example has a considerable amount of control over the rules of the buying and selling game which we all play because of its ability to manipulate the system through taxation and otherwise. What advice should be given to such an individual or institution if the aim of the advice is to change the rules of the game in such a way that the result of its play is an outcome more in accordance with a given criterion of social justice than is the case with the game currently played? This is a question of game *design*. First the games from a particular class of games must be solved and then the game whose solution is optimal with respect to the given social criterion is chosen.

The design of voting systems for legislative assemblies or of voting systems for the election of such assemblies by the population at large properly belong under this heading. However, from the point of view of a game theoretician, a more fascinating class of problem occurs when the prospective players of the game have information that is denied to the game designer before the play of the game. But this information may be essential to the designer if his given objective is to be achieved. He must then seek to build into the game that he designs a mechanism that provides the players with an *incentive* to reveal at least some of the private information that they hold.

The design of auctions provides an example of this type of problem in perhaps its purest form. Usually at auctions successively higher bids are made until nobody is willing to bid any further, in which case the object goes to the highest bidder provided his bid exceeds the reservation price of the seller. At a Dutch auction on the other hand, the auctioneer announces successively lower prices until somebody intervenes with an undertaking to buy at the last price quoted. Sealed bid auctions, as employed in the sale of the Alaskan oil concessions, are another alternative. If the seller has control over the rules of the auction (as in the Alaskan oil concession sale), he might reasonably ask himself what rules are likely to produce a maximum selling price. His difficulty lies in the fact that he does not know the valuations that the potential buyers set upon the object he has to sell and they are unlikely to reveal this information if they can avoid it because each buyer's aim is to secure the objective for a price as far below his valuation as possible.

As an example of what is possible, consider the following type of auction. Each buyer is required to submit a sealed bid. The object is then sold to the highest bidder but for the amount bid by the *second* highest bidder. Rational buyers at this auction will simply bid their true valuations thus revealing their private information. No buyer can benefit from bidding less than his valuation because, whatever the bids the other players may have sealed in their envelopes, such a bid can only decrease his probability of winning without altering the amount he will have to pay if he does win. Equally, no player can benefit from bidding more than his valuation since, if this is necessary in order to win, it must be because some other player has sealed a bid in his envelope which is already at least equal to the valuation of the first player in which case the second highest bid will be at least as large as the maximum the first player is willing to pay and may be larger. Such an auction guarantees the seller the *second* highest valuation as the price of the object he has to sell provided the buyers are rational. He will of course be able to do better in general by using more complicated rules. On the other hand, he may be better advised to keep things fairly simple if he suspects that the buyers are not very thoughtful. (It can be quite profitable, for example, to auction a dollar bill to an audience of economics students according to the usual rules *except* that *every* bidder is required to pay his highest bid even if he does not win.)

It will be evident that the auction example is a paradigm for a large class of problems of very considerable practical importance: notably, taxation, optimal incentive schemes for factor employees, and the writing of insurance policies.

1.4 Emphasis

The game theory literature has been much influenced by the original book of Von Neumann and Morgenstern (1944). Their book splits into two parts of which the first is concerned with games in which the two players have interests which are diametrically opposed (two-person *zero-sum* games). The spirit in which these are studied is what we have called *prescriptive*. Two-person zero-sum games are a useful tool for studying conflict but clearly have little or nothing to offer for the study of cooperation. In the second part of their book, Von Neumann and Morgenstern therefore turn their attention to the question of coalition formation in multiplayer games in which negotiations can result in binding agreements (*cooperative* games). For this problem, they essentially abandon the strategic methodology of the first part of the book and adopt instead a very much more *descriptive* approach.[18] They speak, for example, of "standards of behavior."

The literature by and large has followed this dichotomy of approach. By far the bulk of what has been written has been devoted to developing Von Neumann and Morgenstern's indisputably successful theory of two-person zero-sum games. Indeed, "game theory" has come to *mean* the "theory of two-person zero-sum games" for many people, especially mathematicians. Of the remainder of the literature, most has been devoted to following up or proposing alternatives to the more controversial "descriptive" ideas of the second part of the book.

More recently, the emphasis has changed and, to my mind, very much for the better. The aim of models is now more frequently recognized as *investigative* and attention has turned to a more unified methodology. The theory of two-person zero-sum games has been replaced by a theory of *noncooperative* games in which there may be many players and in which the payoffs need not sum to zero. This work is founded on the pioneering paper of Nash (1951) in which he formulated the basic solution concept of noncooperative theory: the *Nash equilibrium*. Many difficulties remain, especially those which concern the right refinement of Nash equilibrium when questions of information and timing of decisions are crucial. Nevertheless, substantial progress has been made. At the same time, progress has also been made in *cooperative theory* by following up another idea of Nash (1951). His proposal was that one should seek to reduce cooperative theory to noncooperative theory by formalizing the appropriate negotiation procedures for analyzing the result as a noncooperative game. One then need no longer rely on *ad hoc* assumptions about "standards of behavior."

It is this newer tradition which dominates the rest of this book. Its concern is with noncooperative theory and, in particular, with the idea of an equilibrium. Chapters 2 and 3 provide an elementary introduction to the notion of a Nash equilibrium. Chapters 4, 5, and 6 are attempts to explore the idea of an equilibrium at a less superficial level. The aim is to broaden the scope of the available theory by uncovering the implicit assumptions built into current practice. Doubtless, what is written on these topics will be found wanting. It is certainly inadequate in many respects. However, I hope that no one will be left unconvinced of the necessity for thinking about the aspects of human behavior studied and of the immense potential that game theory has for making sense both of ourselves and of our society.

NOTES

1 In the case of the matching pennies game, for example, people are ofte initially unhappy with the notion that optimal reasoning can lead to random behavior. Howard (1971) goes as far as including a version of the matching pennies game in his book *Paradoxes of Rationality.*

2 The general pattern of optimal play was discovered by Von Neumann in the case of a simplified poker model. The particular result quoted here is from Cutler (1975).

3 Only with a Jack-high straight flush should you call!

4 Even less skill, of course, is required to push the nuclear button.

5 See Skinner (1953).

6 An English expression meaning an individual chosen at random.

7 Binmore (1987b) provides some relevant references. The evidence is by no means clear-cut.

8 Güth, Schmittberger, and Schwarze (1982). Binmore, Shaked, and Sutton (1985) obtain similar results for inexperienced subjects. The results in the second paper make it hard to rationalize the behavior as "optimizing with a strange utility function."

9 One could of course say that man is a very clever laboratory rat or a chocolate-dispensing machine of unusually advanced design but this would be a misleading use of the terms "laboratory rat" and "chocolate-dispensing machine."

10 In the sense of Karl Popper (1959). In contradistinction, the stimulus–response model has no explanatory power in this sense because *any* experimental observations will be consistent with the model provided that the stimulus–response mechanism is allowed to be sufficiently complex. This is not to say, however, that the stimulus–response "explanation" is not without value as a means for organizing one's thoughts.

11 In statistical terms.

12 In discussing whether a computer could ever conceivably be regarded as "self-conscious," Turing observed that, if its responses to stimuli and in particular to questioning were indistinguishable from those of a human, then for operational purposes it would be as "self-conscious" as a human in so far as those observing its behavior are concerned. This attitude avoids the necessity of having a view on the "consciousness" issue or indeed on "free-will" and the "soul."

13 This remark is not intended as a shot in the holism versus reductionism war.

Whether one adopts a "holistic" or a "reductionist" viewpoint would appear to depend on what aspects of a problem one wishes to study, in which case it seems absurd to make a once-and-for-all choice between the two.

14 At a Walrasian equilibrium, consumers maximize given their budget constraints. If a new distribution of goods etc. improves the lot of some consumers and harms none, then it necessarily follows that the value of the new total consumption (at Walrasian prices) will exceed the value of the old total consumption. It follows that, since supply must meet demand, the new total production must be valued (at Walrasian prices) above the old total production. But, at a Walrasian equilibrium, producers *maximize* the value of their production, i.e. their profit. However, if the new total production value is higher at least one of the producers must be making a higher profit than before. This is a contradiction.

15 "Zero-sum" and "team" games need to be excluded; also a certain naiveté in interpreting Dahrendorf's statement is necessary.

16 It goes without saying that there is a school of "naive positivist" historians who maintain otherwise and claim that a historian can confine himself to reporting "facts."

17 Except in so far as one might seek to explain the evolution and/or stability of ethical ideas in game-theoretic terms. But this is a separate issue.

18 Although it is not at all clear that Von Neumann and Morgenstern would have accepted this distinction.

REFERENCES

Aumann, R. 1987: "What is game theory trying to accomplish?," in *Frontiers of Economics*, eds Arrow and Honkapohja. Oxford: Blackwell.

Binmore, K. 1987a: "Modeling rational players I," *Economics and Philosophy* 3, 179–214.

Binmore, K. 1987b: "Experimental economics," *European Economic Review* 31, 257–64.

Binmore, K., Shaked, A., and Sutton, J. 1985: "Testing noncooperative game theory: a preliminary study," *American Economic Review* 75, 1178–80.

Cutler, W. 1975: "An optimal strategy for pot-limit poker," *American Mathematical Monthly* 82, 368–76.

Dahrendorf, R. 1958: "Out of Utopia: toward a reorientation of sociological analysis," *American Journal of Sociology* 64, 115–27.

Dawkins, R. 1976: *The Selfish Gene*. Oxford: Oxford University Press.

Debreu, G. 1959: *The Theory of Value*. New York: Wiley.

Güth, W., Schmittberger, R., and Schwarze, B. 1982: "An experimental analysis of ultimatum bargaining," *Journal of Economic Behavior and Organization* 3, 367–88.

Harsanyi, J. 1977: *Rational Behavior and Bargaining Equilibrium in Games and Social Situations*. Cambridge: Cambridge University Press.

Howard, N. 1971: *Paradoxes of Rationality: Theory of Metagames and Political Behavior*. Cambridge, Mass.: MIT Press.

Keynes, J. M. 1936: *The General Theoryof Employment, Interest and Money*. London: Macmillan.
Lakatos, I. 1976: *Proofs and Refutations*, eds Worral and Zahar. Cambridge: Cambridge University Press.
Monod, J. 1972: *Chance and Necessity*. Glasgow: Collins.
Nash, J. 1951: "Noncooperative games." *Annals of Mathematics* 54, 286–95.
Pólya, G. 1962: *Mathematical Discovery*. New York: Wiley.
Popper, K. 1959: *The Logic of Scientific Discovery*. New York: Basic Books.
Rawls, J. 1972: *A Theory of Justice*. Cambridge, Mass.: Harvard University Press.
Shapley, L. 1953: "A value for *n*-person games," in *Contributions to the Theory of Games II* (Annals of Mathematical Studies 24), eds Kuhn and Tucker. Princeton, N.J.: Princeton University Press.
Skinner, B. 1953: *Science and Human Behavior*. New York: Macmillan.
Thorpe, E. 1966: *Beat the Dealer*. New York: Random House.
Von Neumann, J., and Morgenstern, O. 1944: *The Theory of Games and Economic Behavior*. Princeton, N.J.: Princeton University Press.

2 Nash equilibrium

Where infant punks their tender voices try,
And little Maximins the gods defy.

Dryden

2.1 Classifying games

Traditional game theory takes as its basic distinction that between *cooperative* games and *noncooperative* games. In cooperative games, the players are assumed to be free to communicate in any way they choose before and during the game. More importantly, they are also assumed to be able to *bind* themselves to any agreements that may be reached during such bull sessions. In cooperative games, it obviously matters whether players are able to make payments to each other on the side in order to induce agreement to a proposal that would not otherwise be attractive. This leads to a further division of cooperative games into those in which *compensation* of this kind (via the exchange of *side-payments*) is or is not permitted. Following Von Neumann and Morgenstern (1944), it has been usual to analyze cooperative games in which compensation is admitted without an explicit study of the strategic opportunities open to the players.

A noncooperative game ought properly to be defined as a game that is not cooperative. More often, the terminology is used to signify a game in which agreements are never binding on the players. This leaves open the question of the extent to which the players can communicate before and during the game. Such a state of affairs is often unsatisfactory. For example, in bargaining situations it is often crucial to know when offers can be made and for how long they remain open.

The word *contest* will be used to describe a game in which *no* informal communication (explicit or implicit) is allowed between the players before or during the game. This does not exclude the possibility that the players may be able to signal each other by their choice of moves as the game proceeds. Bidding in bridge is a good example. Such *formal* communication, explicitly allowed for in the rules, is not excluded in a contest. What is disbarred is any analysis that presupposes *informal* communication possibilities that are not written into the rules. In bridge, this would include idle (or not-so-idle) chatter during the play or, more significantly, pre-play meetings to discuss bidding systems and tactics. Since such pre-play meetings matter a lot in bridge, it would therefore be less than realistic to analyze the game of bridge as a contest.

The Nash (1951) program calls for *all* games to be reduced to contests. In principle, this aim can be achieved by incorporating the informal opportunities available to the players for communication and negotiation as *formal* moves in a game with an *expanded* rule-book. The game that results from using this expanded rule-book then has to be treated as a contest because there is nothing left that the players might do that is not already taken account of in the rules. In most cases, however, the Nash program remains only an ideal at which to aspire. One has only to consider the difficulties involved in inventing formal rules that would be adequate to codify anything relevant that might take place at a pre-play bridge meeting to see why this is so. Nevertheless, contests remain the fundamental type of noncooperative game.

A different distinction is that between *normal-form* games and *extensive-form* games. Von Neumann and Morgenstern (1944) took the view that all games can be reduced to normal form *without loss of generality*. Their argument is nowadays seen as depending on assumptions concerning the ability of players to make pre-commitments about their future behavior that are not realistic.[1] More neutral terms will therefore be substituted and the distinction drawn will be between *static* games and *dynamic* games. A static game is one which can be thought of as taking place at a single instant of time. All the action is therefore simultaneous. A dynamic game will simply be taken to be a game that is not static. An adequate description of such a game therefore requires at least two points in time, and players are forced to condition their behavior both on what has happened in the past and on what might happen in the future.

Yet another distinction is concerned with describing the possible outcomes of the game. A game will be said to be *strictly competitive* if no player ever ranks one possible outcome above another without at least one other player being in disagreement about the ranking. (The idea is only loosely related to the notion of "perfect competition" as used by economists.) Most games, of course, are not strictly competitive and, in any case, it is only for two-player games that the idea really bites. In such games, the interests of the players will then be diametrically opposed. In games with three or more players, there remains room for coordination between some of the players for the purposes of exploiting the others.

This chapter is largely concerned with static games. The games are always treated as contests. This is a harmless requirement for two-player strictly competitive games because, in such games, the players have nothing worthwhile to discuss. In chess, for example, Black may try to persuade White to surrender his Queen but White is unlikely to be convinced since, if this is good for Black, it cannot simultaneously be good for White. For this reason, nearly all the examples in this chapter are of two-person *zero-sum* games as studied by Von Neumann and Morgenstern (1944). Such games are always the first to be studied in traditional game theory books and hence provide a convenient anchoring ground for the terminology of the theory.

2.2 Static games

Formally, a static n-person game is defined by n *strategy spaces* $S_1, S_2, \dots S_n$ and n *payoff functions* $\phi_i: S_1 \times S_2 \times \dots \times S_n \to \mathbb{R}$ $(i = 1, 2, \dots, n)$. We interpret $\phi_i(s_1, s_2, \dots, s_n)$ as the utility which player i will derive from the *strategy profile* $(s_1, s_2, \dots, s_n)$ obtained when player I chooses strategy s_1, player II chooses strategy s_2, and so on.

A two-person game in which player I's strategy space P_1 contains m elements and player II's strategy space P_2 contains n elements will be called an $m \times n$ *bimatrix* game. The reason is that the payoff functions can be specified with two $m \times n$ matrices. Figure 2.1 provides a 2×3 example. It is useful to think of player I's strategy space P_1 as constituting the rows of the matrices and player II's strategy space P_2 as constituting the columns.

In the example, if player I chooses his first strategy (row 1) and player II chooses her third strategy (column 3), then the payoff to player I is 3 and the payoff to player II is 2. In this particular example, the payoff matrix for player I is A and that for player II is B, where

$$A = \begin{pmatrix} 1 & 2 & 3 \\ 3 & 1 & 2 \end{pmatrix} \qquad B = \begin{pmatrix} 3 & 1 & 2 \\ 2 & 3 & 1 \end{pmatrix}$$

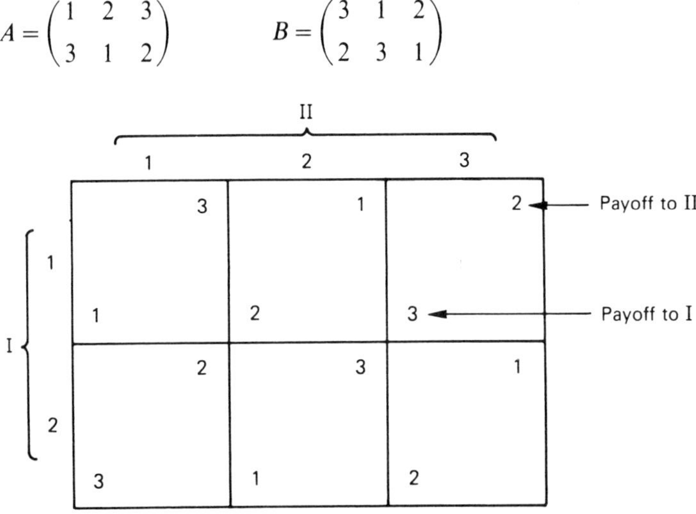

Figure 2.1

The convention that odd players are male and even players are female is adopted only partly for reasons of sentiment. Without such a convention, some sentences have to be written without pronouns for fear of introducing ambiguities and can be very painful to read.

In analyzing a static game we proceed as though the moves were made simultaneously, but this is just for convenience of expression. A bimatrix game, for example, equally well represents the situation in which player I moves first and player II moves second, *provided* that player II is ignorant of player I's move when she decides what to do. Figure 2.2(a) illustrates how this arrangement may be represented with the aid of a *game tree.* Figure 2.2(b) shows the (strategically equivalent) arrangement in which it is player II who moves first and player I who moves second in ignorance of player II's move.

The game begins at the initial node of the tree (represented by ○ in the diagrams). The information sets indicate what the players know about the past history of the game when they make a decision. Thus, in figure 2.2(a), player I knows that he is making the first move of the game when he makes a selection from his choice of $\{1,2\}$. If he chooses 2 from this set, then we move to the right-hand node in player II's information set. But player II does not know this. Her information set indicates that, when she makes a selection from her choice set $\{1,2,3\}$, she knows only that player I has played either 1 or 2 but not which of these. She would, of course, *like* to know what player I has played. If she knew that player I had player 2, then her optimal response would be to play 2 herself (which yields a payoff of 3 to her and 1

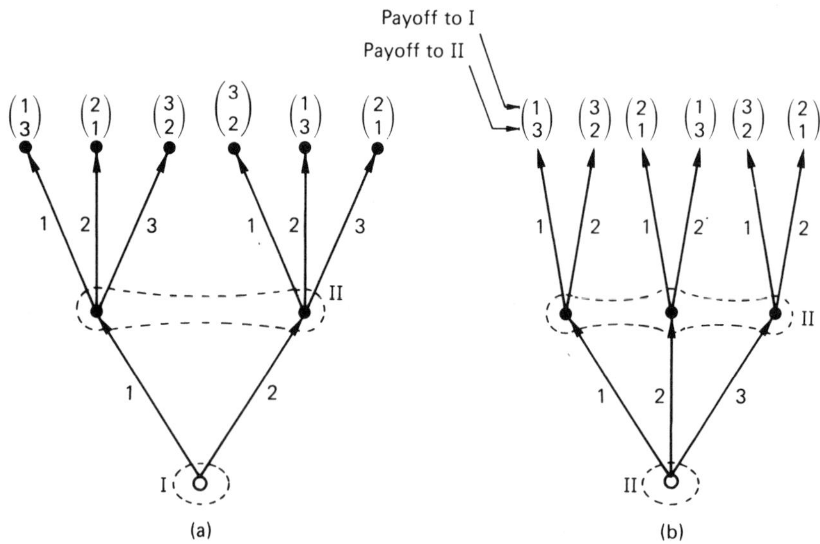

Figure 2.2

to player I). But she has to take into account the fact that player I *might* have played 1 or, as we shall see, that player I might have chosen 1 with probability $1-p$ and 2 with probability p. The fact that she chooses in ignorance of player I's choice therefore leaves her in the same position as if both chose simultaneously.

The rows and columns in a bimatrix game are called *pure strategies* to distinguish them from *mixed strategies*. A player is said to use a mixed strategy if he or she employs a random device (such as a roulette wheel) in deciding what to do in a game. The nature of the random device is irrelevant beyond the probabilities which it assigns to each of the player's pure strategies. This makes it possible to describe mixed strategies using a simple vector notation. For example, a mixed strategy for player II in the game of figure 2.1 is a 3×1 column vector $\mathbf{q}$ in which the ith coordinate q_i represents the probability with which player II will choose her ith pure strategy (i.e. her ith column). If player I uses mixed strategy $\mathbf{p}$ and player II uses mixed strategy $\mathbf{q}$, then the *expected* payoff to player I is

$$\mathbf{x} = \mathbf{p}^{\mathrm{T}} A \mathbf{q} = (p_1, p_2) \begin{pmatrix} 1 & 2 & 3 \\ 3 & 1 & 2 \end{pmatrix} \begin{pmatrix} q_1 \\ q_2 \\ q_3 \end{pmatrix}$$

provided that the game is to be analyzed as a *contest*, i.e. without pre-play interaction between the players. This assumption is invoked in our implicit use of the fact that the random devices employed by the players are *independent*[2] and hence, for example, the probability that the final outcome will be (row 1, column 3) is $p_1 \times q_3$. The *expected* payoff to player II under similar conditions is

$$\mathbf{y} = \mathbf{p}^{\mathrm{T}} B \mathbf{q} = (p_1, p_2) \begin{pmatrix} 3 & 2 & 1 \\ 2 & 3 & 1 \end{pmatrix} \begin{pmatrix} q_1 \\ q_2 \\ q_3 \end{pmatrix}$$

The computation of these expected payoffs would be of little significance if we were not entitled to assume that the players were interested in maximizing them. This makes it important that the entries in the matrices be interpreted as *Von Neumann and Morgenstern utilities* for the possible outcomes of the game. Not the least of Von Neumann and Morgenstern's (1944) achievements was to demonstrate that any "rational" player will act *as though* he or she were seeking to maximize the expected value of a certain function $\phi\colon \Omega \to \mathbb{R}$ defined over the set Ω of possible outcomes. Such a function ϕ is called a Von Neumann and Morgenstern utility function for the player in question.

At first sight, it seems paradoxical that mixed strategies might be used by rational players. However, it is easy to agree that we can do rational players no harm by offering them the opportunity to use mixed strategies if they like. In our bimatrix example, this involves enlarging player I's strategy space from $P_1 = \{\text{row 1, row 2}\}$ to the space M_1 of all his mixed strategies. Similarly, player II's strategy space is enlarged from $P_2 = \{\text{column 1, column 2, column 3}\}$ to the space M_2 of all her mixed strategies. Mathematically, it is convenient to begin by identifying player I's pure strategies, row 1 and row 2, with the vectors $\mathbf{e}_1^1 = (1, 0)^{\mathrm{T}}$ and $\mathbf{e}_2^1 = (0, 1)^{\mathrm{T}}$. After this identification,

$$M_1 = \{\mathbf{p}\colon p_1 \geqslant 0,\ p_2 \geqslant 0, \text{ and } p_1 + p_2 = 1\}$$

is just the convex hull of P_1. In the same way, player II's pure strategies are identified with $\mathbf{e}_1^2 = (1, 0, 0)^{\mathrm{T}}$, $\mathbf{e}_2 = (0, 1, 0)^{\mathrm{T}}$, and $e_3^2 = (0, 0, 1)^{\mathrm{T}}$, so that

$$M_2 = \{\mathbf{q}\colon q_1 \geqslant 0,\ q_2 \geqslant 0,\ q_3 \geqslant 0, \text{ and } q_1 + q_2 + q_3 = 1\}$$

is the convex hull of P_2. In the enlarged game, with strategy spaces M_1 and M_2, the payoff functions are $\phi_1\colon M_1 \times M_2 \to \mathbb{R}$ and $\phi_2\colon M_1 \times M_2 \to \mathbb{R}$, where

$$\phi_1(\mathbf{p}, \mathbf{q}) = \sum_{i=1}^{2} \sum_{j=1}^{3} \phi_1(\mathbf{e}_i^1, \mathbf{e}_j^2) p_i q_j = \mathbf{p}^{\mathrm{T}} A\mathbf{q}$$

$$\phi_2(\mathbf{p}, \mathbf{q}) = \sum_{i=1}^{2} \sum_{j=1}^{3} \phi_2(\mathbf{e}_i^1, \mathbf{e}_j^2) p_i q_j = \mathbf{p}^{\mathrm{T}} B\mathbf{q}$$

Sometimes it is adequate, in analyzing a game, to work only with pure strategies, in which case we just take $S_1 = P_1$ and $S_2 = P_2$ in our definition of a static game. On other occasions, it is necessary to turn to mixed strategies and to take $S_1 = M_1$ and $S_2 = M_2$.

A little care is worth taking when introducing mixed strategies. It would be easy, for example, to make the error of supposing that a two-person strictly competitive game is just the same as a two-person zero-sum game.

A *constant-sum* game is a game in which the payoffs always sum to a constant, i.e. there exists a constant c such that

$$\phi_1(s) + \phi_2(s) + \ldots + \phi_n(s) = c$$

for all $s \in S_1 \times S_2 \times \ldots \times S_n$. In a two-person zero-sum game,

$$\phi_2(s) = -\phi_1(s)$$

for all $s \in S_1 \times S_2$. In a two-person strictly competitive game, we have instead that, for all $s \in S_1 \times S_2$ and $t \in S_1 \times S_2$,

$$\phi_1(s) < \phi_1(t) \Leftrightarrow \phi_2(s) > \phi_2(t)$$

This last equivalence tells us that $-\phi_1$ is a utility function for player II on $S_1 \times S_2$. This means that she prefers s to t if and only if $-\phi_1(s) > -\phi_1(t)$. So why not replace ϕ_2 by $-\phi_1$ and obtain a zero-sum game?

As long as we stick to the strategy spaces S_1 and S_2, there is no reason at all not to adopt this simplification. But what if we begin with spaces P_1 and P_2 of pure strategies and then enlarge these to spaces M_1 and M_2 of mixed strategies? We have seen how to proceed when ϕ_1 and ϕ_2 are Von Neumann and Morgenstern utility functions. But no guarantee exists that $-\phi_1$ is a Von Neumann and Morgenstern utility function for player II.[3] If players I and II have different attitudes to the taking of risks, then it certainly will not be true.

This point is sometimes lost sight of when games like poker or backgammon with an intrinsic chance element are treated automatically as zero-sum. It is true that, what one player wins, the other player (or players) necessarily loses. It is also reasonable to model the players as preferring more money to less. The game is therefore strictly competitive as regards deterministic outcomes. But chance moves and mixed strategies also matter. To treat the game as strictly competitive when these are also admitted is to make an assumption that may well not be at all innocent.

2.3 Zero-sum matrix games

We have seen that, if each player has a finite set of pure strategies, then we can represent the payoff functions by a pair of matrices A and B. An example of a zero-sum 2×2 matrix game is illustrated in figure 2.3(a). Since B is necessarily $-A$ in zero-sum games, it is redundant to write B down as well as A. It is therefore usual to specify a finite two-person zero-sum game with a single matrix as indicated in figure 2.3(b). The entries are then gains to player I but *losses* to player II.

We continue with a sequence of examples. These are included largely as an excuse to introduce a number of definitions in an informal manner.

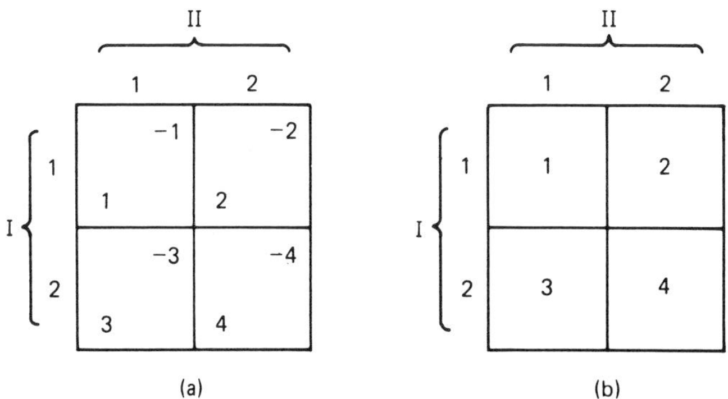

Figure 2.3

Example 2.3.1

In this game, player I's second pure strategy (row 2) *strongly dominates* his first pure strategy (row 1). Mathematically, this means that "1 < 3 and 2 < 4," which we write as

$$(1, 2) \ll (3, 4)$$

(In general, if $\mathbf{u}$ and $\mathbf{v}$ are n-dimensional vectors, $\mathbf{u} \ll \mathbf{v}$ means that $u_i < v_i$ $(i = 1, 2, \ldots, n)$ and $\mathbf{u} \leqslant \mathbf{v}$ means $u_i \leqslant v_i$ $(i = 1, 2, \ldots, n)$.) The game-theoretic interpretation of the fact that player I's second pure strategy strongly dominates his first is that, *whatever* player II does, player I will necessarily get a better payoff from his second pure strategy than from his first. Hence a rational player I will *always* choose his second pure strategy.

Similarly, player II's first pure strategy strongly dominates her second because $(1, 2)^T \ll (2, 3)^T$. (Recall that the matrix entries are *negative* payoffs for player II.) Thus a rational player II will *always* choose her first pure strategy.

These domination arguments show that the solution outcome of the game is the pure strategy profile (row 2, column 1). This yields a payoff of 3 to player I and -3 to player II.

Example 2.3.2

The game of figure 2.4(a) is marginally more subtle than the preceding example because player II does not have a dominated strategy. However, player II can predict, with certainty, that a rational player I will choose his second pure strategy because this strongly dominates his first. Since row 1 will not be used, player II is then faced with the "reduced game" of figure 2.4(b). In this reduced game, player II's second pure strategy strongly

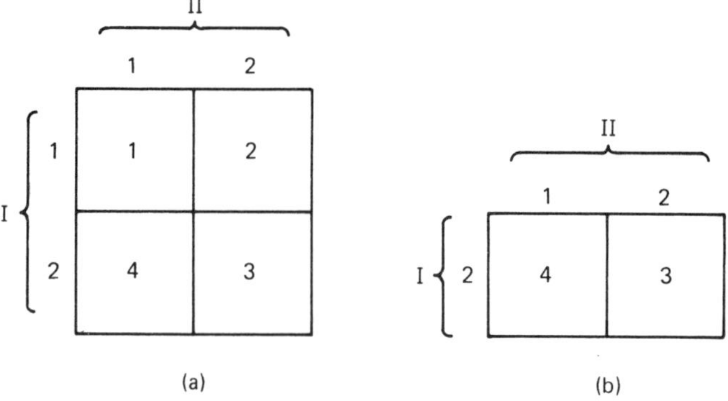

Figure 2.4

dominates her first. The solution of the game is therefore the pure strategy profile (row 2, column 2). This again yields a payoff of 3 to player I and -3 to player II.

Note that this argument requires, not only that players I and II are rational to the extent that they do not play strongly dominated strategies, but also that player II *knows* that I is rational to this extent.

Sometimes, following a suggestion of Luce and Raiffa (1957), it is possible to solve a game by deleting strongly dominated strategies over and over again. In such cases, it may be necessary to assume statements like "Player II knows that I knows that II knows that I knows that II will not choose a strongly dominated strategy." Game theorists sometimes express the fact that the totality of such assumptions is to be made with a sentence of the form, "It is *common knowledge* that strongly dominated strategies will not be used." (For a rigorous definition of "common knowledge," see Aumann (1976) and chapter 4.) Without such common knowledge hypotheses, it would be impossible to rely on chains of reasoning which begin, "If I think that he thinks that I think ...." But such chains of reasoning always lie at the heart of traditional game-theoretic arguments even though this fact is not always made explicit.

The extent to which game-theoretic conclusions depend on such heavy common knowledge assumptions is often not appreciated. But where the necessary common knowledge is absent the conclusions cannot reasonably be expected to apply. The results of attempts to apply the theory under these circumstances are therefore usually disappointing. A theory which was not subject to such heavy constraints would obviously be more satisfactory. However, just as bricks cannot be made without straw, so conclusions cannot be obtained without hypotheses to deduce them from.

Example 2.3.3

In the game of figure 2.5(a), player I's second pure strategy *weakly dominates* his first. We have that $(1, 2, 3) < (3, 4, 3)$. (The notation $\mathbf{u} < \mathbf{v}$ means that $\mathbf{u} \leqslant \mathbf{v}$ but $\mathbf{u} \neq \mathbf{v}$.) Player II's first pure strategy strongly dominates her second but only weakly dominates her third. Neither player can suffer from deleting a weakly dominated strategy from consideration. We may therefore conclude that the pure strategy profile (row 2, column 1) is *a* solution of the game.

However, nothing guarantees that games have a *unique* solution any more than quadratic equations. If player II argues that player I will not choose his first pure strategy because this is weakly dominated and decides not to choose her second pure strategy because this is strongly dominated, then we obtain the reduced game of figure 2.5(b). In this reduced game, player II is indifferent between her pure strategies 1 and 3 and may choose either (or a probabilistic mixture of the two). In particular, the considerations we have

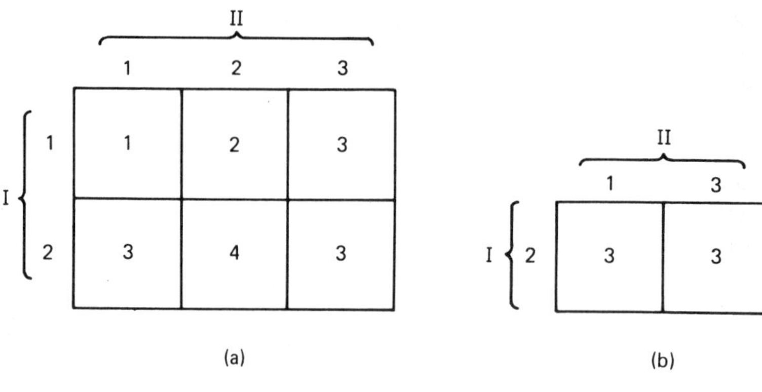

Figure 2.5

discussed so far leave it open that the pure strategy profile (row 2, column 3) may be a solution. (We return to this point in section 2.9.)

Example 2.3.4

In the game of figure 2.6(a), no pure strategies are dominated, but the pure strategy profile (row 2, column 1) is a *saddle point* of the matrix. In game theory, this means that the entry 2 is simultaneously smallest in its row and largest in its column. In consequence, player I's pure strategy 2 is an *optimal response* (or a best *reply*) to a choice of pure strategy 1 by player II and,

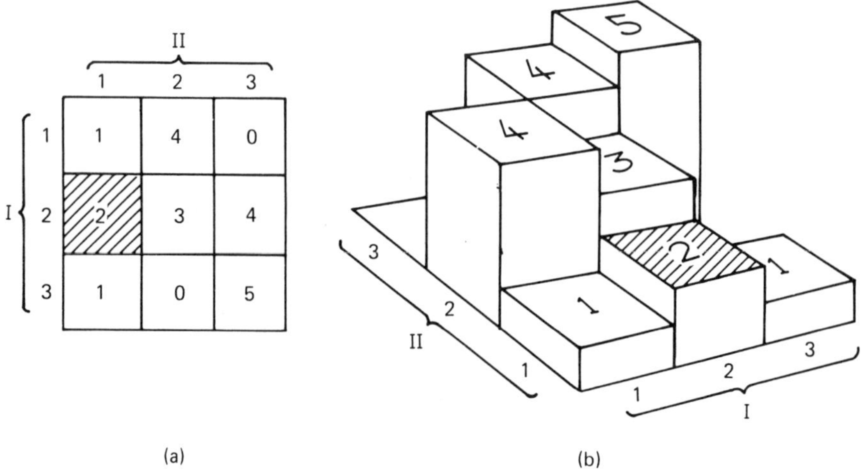

Figure 2.6

simultaneously, player II's pure strategy is an *optimal response* to a choice of pure strategy 2 by player I.

A strategy pair (s,t) in which s is an optimal response to t and t is an optimal response to s is called a *Nash equilibrium* for a two-person game. Thus (row 2, column 1) is a Nash equilibrium in the current example. In section 2.5, a detailed explanation is given of why attention should be focused on Nash equilibria in static contests. For the moment, we observe simply that any strategy pair without this property is "self-destabilizing" in the sense that at least one player would not follow a recommendation to play according to such a strategy pair if he or she thought the other player *would* follow the recommendation.

Example 2.3.5

In the game of figure 2.7(a), there are no dominated strategies and no saddle points. The latter consideration implies that the game has *no* Nash equilibria in pure strategies. But a Nash equilibrium does exist provided that mixed strategies are allowed.

Suppose that player II chooses mixed strategy $(1-q, q)^{\mathrm{T}}$. This means that she uses her first pure strategy with probability $1-q$ and her second with probability q. The respective expected payoffs to player I from his first and

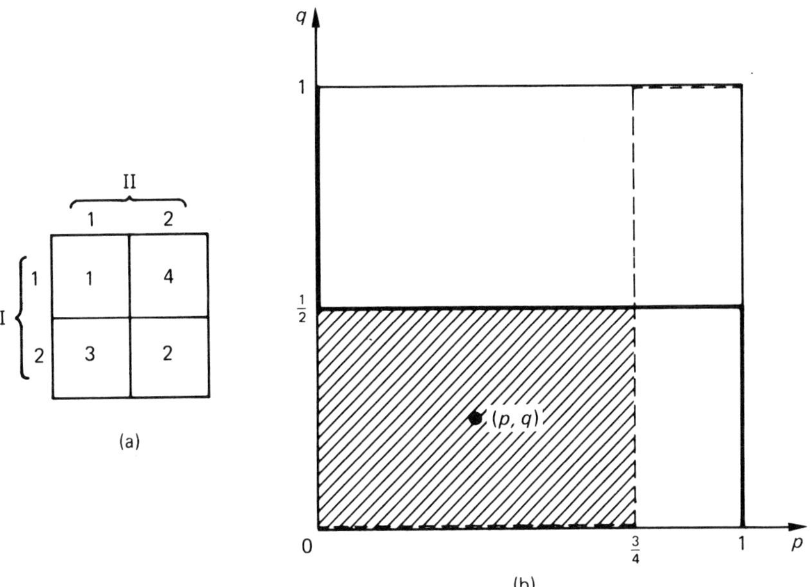

Figure 2.7

second pure strategies are therefore

$$E_1 = (1-q) + 4q = 1 + 3q$$

and

$$E_2 = 3(1-q) + 2q = 3 - q$$

When is $(1-p, p)^{\mathrm{T}}$ an optimal response by player I to the choice of $(1-q, q)^{\mathrm{T}}$ by player II? If $E_1 > E_2$ (which means that $q > 1/2$), then player I will want to use his first pure strategy and so $p = 0$. If $E_1 < E_2$ (which means that $q < 1/2$), then the optimal response is $p = 1$. If $E_1 = E_2$ (which means that $q = 1/2$), then any response is equally good and so any p with $0 \leqslant p \leqslant 1$ is an optimal response. A similar analysis shows that $q = 0$ is the optimal response for player II when $p < 3/4$, $q = 1$ is the optimal response when $p > 3/4$, and, when $p = 3/4$, any q with $0 \leqslant q \leqslant 1$ is optimal. These conclusions are graphed in figure 2.7(b). Player II's optimal response curve is drawn with a broken line to distinguish it from player I's. For example, when (p, q) is in the shaded region, it is optimal for player I to react to q by playing his second pure strategy and for player II to react to p by playing her first pure strategy.

The optimal response curves cut where $p = 3/4$ and $q = 1/2$. This means that $p = 3/4$ is an optimal response to $q = 1/2$ and $q = 1/2$ is an optimal response to $p = 3/4$. It is therefore a Nash equilibrium for player I to use his second pure strategy with probability 3/4 and for player I to use his second pure strategy with probability 1/2. Player I then gets an expected payoff of 5/2 and player II an expected payoff of $-5/2$. No other Nash equilibria exist because the optimal response curves cut in only one point.

In this example, each mixed strategy $(1-r, r)^{\mathrm{T}}$ can be identified with a number r in the compact interval $[0, 1]$. We can therefore view the game as a static game with strategy spaces $S_1 = S_2 = [0, 1]$ and payoff functions $\phi_1 = -\phi_2 = \phi$, where ϕ: $[0, 1] \times [0, 1] \to \mathbb{R}$ is defined by

$$x = \phi(p, q) = (1-p, p)\begin{pmatrix} 1 & 4 \\ 3 & 2 \end{pmatrix}\begin{pmatrix} 1-q \\ q \end{pmatrix}$$

Figure 2.8(a) illustrates that $(p, q) = (3/4, 1/2)$ is a saddle point for the function ϕ. Since ϕ is a rather special function (it is bilinear in p and q, and so its graph is a ruled surface), we also illustrate a saddle point $(\tilde{p}, \tilde{q})$ in figure 2.8(b) for a more general function f. The requirement is that $f(p, \tilde{q}) \leqslant f(\tilde{p}, \tilde{q}) \leqslant f(\tilde{p}, q)$ for all admissible p and q.

Example 2.3.6

Various methods exist for computing the Nash equilibria of $m \times n$ matrix

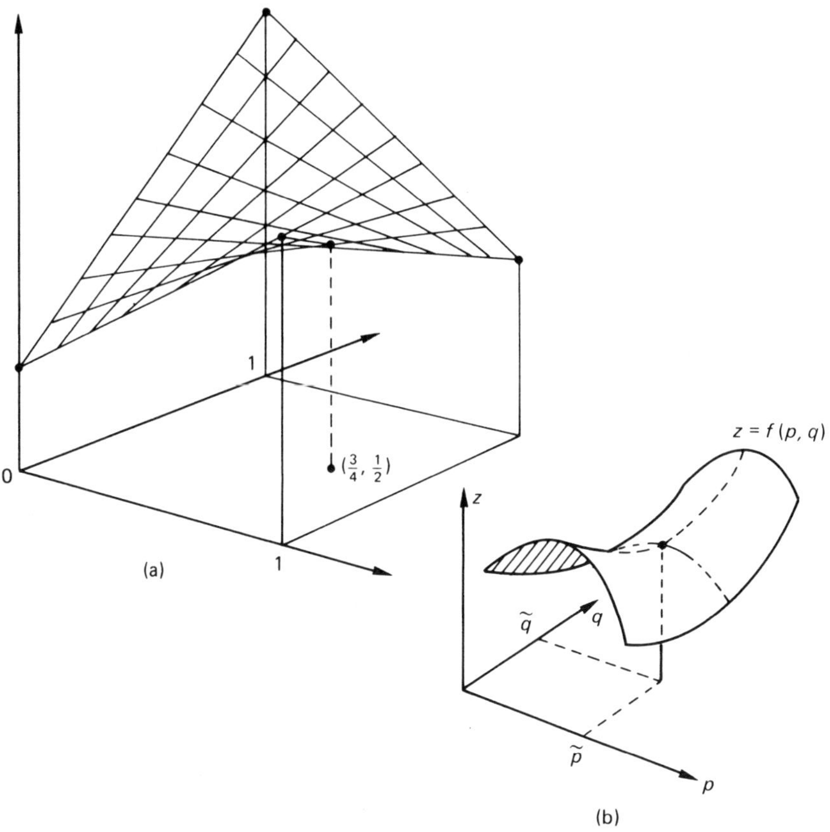

Figure 2.8

games. For $m > 3$ and $n > 3$, linear programming is the technique usually recommended. The book by Owen (1982) is a good reference for this material. The game of figure 2.9 is an example of a 4×4 matrix game for which a Nash equilibrium can be located without such exertions.

If player II uses a mixed strategy which assigns probability $1/4$ to each of her pure strategies, then player I gets an expected payoff of $(1+2+3+4)/4 = 5/2$ whatever he does. Anything is therefore an optimal response for him: in particular, the mixed strategy in which he assigns $1/4$ to each of his pure strategies is an optimal response. A similar argument applies with the roles of the players reversed. It is therefore a Nash equilibrium for each player to choose each of his or her strategies with equal probabilities.

These strategies are examples of *completely* mixed strategies. Such mixed strategies assign a positive probability to *every* pure strategy. A bimatrix game can have at most one Nash equilibrium in completely mixed strategies

		II			
		1	2	3	4
I	1	1	2	3	4
	2	4	1	2	3
	3	3	4	1	2
	4	2	3	4	1

Figure 2.9

but it may have many Nash equilibria in which one or more pure strategies are assigned zero probability.

Example 2.3.7

We now describe what is perhaps the simplest technique for finding all Nash equilibria $(\tilde{p}, \tilde{q})$ of any zero-sum game based on a $2 \times n$ matrix A.

Let H be the set of all column vectors of the form $A\mathbf{q}$, where $\mathbf{q}$ is a mixed strategy for player II. Geometrically, H is simply the convex hull of the columns of A. Figure 2.10 indicates the set H in the particular case when

$$A = \begin{pmatrix} -1 & 0 & 1 & 3 & 5 & 4 \\ 7 & 5 & 3 & 1 & 6 & 4 \end{pmatrix}$$

We next choose v to be the *smallest* real number such that the set $K = \{\mathbf{x}: x_1 \leqslant v \text{ and } x_2 \leqslant v\}$ has at least one point in common with H. For the matrix A with which we are concerned, $v = 2$ as indicated in figure 2.10.

Choose any $\tilde{\mathbf{q}}$ with the property that $A\tilde{\mathbf{q}} \in H \cap K$. In figure 2.10, $H \cap K$ consists of the single point $(2, 2)^T$. Since this point lies midway between $(1, 3)^T$ and $(3, 1)^T$, algebra is not necessary to compute $\tilde{\mathbf{q}}$. In expressing $(2, 2)^T$ as a convex combination of the columns of A, weights of $1/2$ each need to be attached to $(1, 3)^T$ and $(3, 1)^T$ and weights of zero to the other columns. Thus $\tilde{\mathbf{q}} = (0, 0, 1/2, 1/2, 0, 0)^T$.

The next step is to locate a line $\tilde{\mathbf{p}}^T\mathbf{x} = v$ which separates the convex sets H and K. This means that H lies on one side of the line and K lies on the

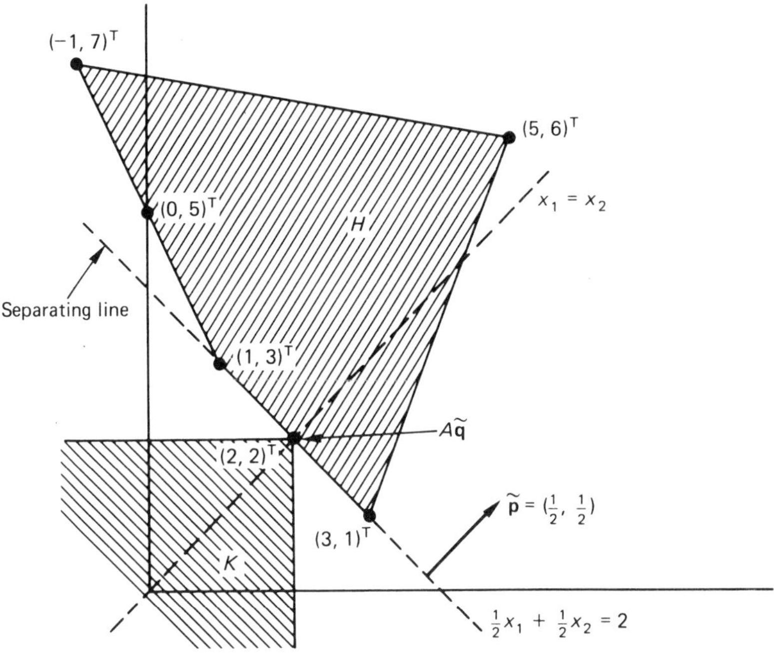

Figure 2.10

other side. Algebraically, we require that each point $\mathbf{h} \in H$ satisfies $\mathbf{p}^{\mathrm{T}}\mathbf{h} \geqslant v$ and that each point $\mathbf{k} \in K$ satisfies $\tilde{\mathbf{p}}^{\mathrm{T}}\mathbf{k} \leqslant v$. Although such sophistication is unnecessary in two-dimensional situations, the existence of a separating line is guaranteed by the theorem of the separating hyperplane. Geometrically, $\tilde{\mathbf{p}}$ is a vector orthogonal to the separating line.

In figure 2.10, the separating line has equation

$$\tfrac{1}{2}x_1 + \tfrac{1}{2}x_2 = 2$$

and so $\tilde{\mathbf{p}} = (1/2, 1/2)^{\mathrm{T}}$. Notice that $\tilde{\mathbf{p}}$ is a mixed strategy for player I.

It remains to explain why the pair $(\tilde{\mathbf{p}}, \tilde{\mathbf{q}})$ is a Nash equilibrium. Observe first that $\tilde{\mathbf{p}}^{\mathrm{T}}\mathbf{h} \geqslant v$ for all $\mathbf{h} \in H$. It follows that $\tilde{\mathbf{p}}^{\mathrm{T}}A\mathbf{q} \geqslant v$ for all player II's mixed strategies $\mathbf{q}$. Secondly, we have that $A\tilde{\mathbf{q}} \in H \cap K$ and so $A\tilde{\mathbf{q}} \in K$. Thus both coordinates of the column vector $A\tilde{\mathbf{q}}$ are no greater than v. Hence no convex combination of these coordinates is greater than v. We deduce that $\mathbf{p}^{\mathrm{T}}A\tilde{\mathbf{q}} \leqslant v$ for all player I's mixed strategies $\mathbf{p}$. A consequence of these results is that $v = \tilde{\mathbf{p}}^{\mathrm{T}}A\tilde{\mathbf{q}}$. Moreover, $\tilde{\mathbf{p}}$ is a best response to $\tilde{\mathbf{q}}$ because

$$\tilde{\mathbf{p}}^{\mathrm{T}}A\tilde{\mathbf{q}} = v \geqslant \mathbf{p}^{\mathrm{T}}A\tilde{\mathbf{q}}$$

for all $\mathbf{p}$, and $\tilde{\mathbf{q}}$ is a best response to $\tilde{\mathbf{p}}$ because

$$-\tilde{\mathbf{p}}^{\mathrm{T}}A\tilde{\mathbf{q}} = -v \geqslant -\tilde{\mathbf{p}}^{\mathrm{T}}A\mathbf{q}$$

for all $\mathbf{q}$. Thus $(\tilde{\mathbf{p}}, \tilde{\mathbf{q}})$ is a Nash equilibrium.

Sometimes there may be many Nash equilibria. The method we have described finds them all. Each $\tilde{\mathbf{q}}$ corresponds to a point in $H \cap K$ and each $\tilde{\mathbf{p}}$ to a line which separates H and K. In the particular case we considered here, $H \cap K$ contains only a single point and only one separating line exists. Thus there is a unique Nash equilibrium.

2.4 Nash equilibrium

Let $S = S_1 \times S_2 \times \ldots \times S_n$. A *Nash equilibrium* for the static contest with strategy spaces $S_1, S_2, \ldots, S_n$ and payoff functions $\phi_i: S \to \mathbb{R}$ $(i = 1, 2, \ldots, n)$ is a strategy profile $\tilde{s} \in S$ for which

$$\phi_1(\tilde{s}_1, \tilde{s}_2, \ldots, \tilde{s}_n) \geqslant \phi_1(s_1, \tilde{s}_2, \ldots, \tilde{s}_n)$$

$$\phi_2(\tilde{s}_1, \tilde{s}_2, \ldots, \tilde{s}_n) \geqslant \phi_2(\tilde{s}_1, s_2, \ldots, \tilde{s}_n)$$

$$\ldots$$

$$\phi_n(\tilde{s}_1, \tilde{s}_2, \ldots, \tilde{s}_n) \geqslant \phi_n(\tilde{s}_1, \tilde{s}_2, \ldots, s_n)$$

for all $s_1 \in S_1, s_2 \in S_2, \ldots, s_n \in S_n$. These inequalities assert that each strategy $\tilde{s}_i$ $(i = 1, 2, \ldots, n)$ is an *optimal response* (or a *best reply*) for player i to the strategies specified by $\tilde{s}$ for the other players.

Nash (1951) proved a version of the theorem stated below.

NASH'S THEOREM

Suppose that $S_1, S_2, \ldots, S_n$ are compact and convex. Suppose that the payoffs $\phi_i(s_1, s_2, \ldots, s_n)$ $(i = 1, 2, \ldots, n)$ are concave functions of each variable s_j separately and continuous functions of $(s_1, s_2, \ldots, s_n)$. Then at least one Nash equilibrium exists.

COROLLARY

Every finite game has at least one Nash equilibrium provided that mixed strategies are admitted.

Nash actually proved the corollary about finite games (finite in that the number of players and the number of strategies is finite). The payoffs for a finite game into which mixed strategies have been introduced are continuous and, not only are they concave in each variable separately, they are linear in each variable separately (section 2.2). The corollary is therefore an immediate special case of the theorem.

Aside from the conditions under which Nash equilibria exist, note should also be taken of the following elementary properties of Nash equilibria.

1 No nash equilibrium every assigns a positive probability to a strictly dominated strategy.
2 Given any weakly dominated strategy, there exists a Nash equilibrium which assigns it zero probability (provided Nash equilibria exist at all).

Example 2.4.1

Consider the two person zero-sum game with matrix

$$A = \begin{pmatrix} 1 & 3 & 2 & 2 \\ 2 & 2 & 2 & 3 \\ -1 & 2 & 1 & 0 \end{pmatrix}$$

Note that player II's first pure strategy *strongly* dominates her fourth. In seeking Nash equilibria, we may therefore delete the fourth column and study only the game that remains, namely

$$A_1 = \begin{pmatrix} 1 & 3 & 2 \\ 2 & 2 & 2 \\ -1 & 2 & 1 \end{pmatrix}$$

In this game, player I's first pure strategy *strongly* dominates his third. Deleting the third row yields

$$A_2 = \begin{pmatrix} 1 & 3 & 2 \\ 2 & 2 & 2 \end{pmatrix}$$

In this game, player II's first pure strategy *weakly* dominates her other strategies. After these have been deleted, player I's second pure strategy *strongly* dominates his first.

These considerations lead to the Nash equilibrium (row 2, column 2) in the original game. It is easy to check that this is a saddle point of the matrix A.

But a little care is necessary since other Nash equilibria exist. We deleted dominated strategies in four successive stages but, at the third, the domination was only *weak*. The other Nash equilibria were lost at this stage. To see this, we return to the matrix A_2 and use the method of example 2.3.7. The set H becomes the line segment joining $(1, 2)^T$ and $(3, 2)^T$ as illustrated in figure 2.11. The number v is 2 and the separating line is $x_2 = 2$. There is therefore a unique $\tilde{\mathbf{p}}_2$, namely $\tilde{\mathbf{p}}_2 = (0, 1)^T$, but any $\tilde{\mathbf{q}}_2$ which places $A_2\tilde{\mathbf{q}}_2$ on the line segment joining $(1, 2)^T$ and $(2, 2)^T$ (i.e. $H \cap K$) will suffice. The mixed

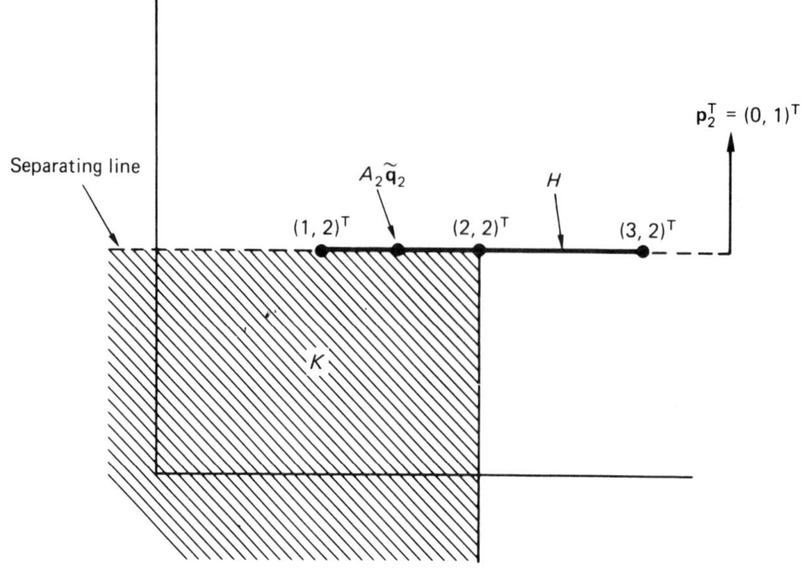

Figure 2.11

strategies with this property are simply those which give more weight to the column $(1, 2)^T$ than to the column $(3, 2)^T$ in the matrix A_2.

Returning to the original game, it follows that the Nash equilibria are those $(\tilde{\mathbf{p}}, \tilde{\mathbf{q}})$ with $\tilde{\mathbf{p}} = (0, 1, 0)^T$ and $\tilde{\mathbf{q}} = (q_1, 1 - q_1 - q_3, q_3, 0)$ where $0 \leqslant q_3 \leqslant q_1 \leqslant 1$ and $0 \leqslant 1 - q_1 - q_3 \leqslant 1$.

2.5 Justification of Nash equilibria

Why should the solution of a static contest necessarily be a Nash equilibrium? There is much scope for confusion on this issue, chiefly because the question has more than one answer depending on how the word "solution" is interpreted.

The traditional interpretation, as given by Von Neumann and Morgenstern (1944), can be explained in the following way. Imagine that an absolutely authoritative game theory book was to be found on the bookshelves of all those ever likely to play a game. What could be said about what would be written in this book? In the case of a static contest, any recommendation would need to specify a Nash equilibrium. Given that the book is to be absolutely authoritative, all players will wish to follow its recommendations. But, if the book does not recommend a Nash equilibrium, then at least one player is being advised to choose a strategy which is not an optimal response to the strategies recommended for the others. At least one player will therefore wish to deviate from the recommendation made to him (unless he believes that other players will deviate). In either case, the book will fail to be authoritative.

This argument depends heavily on it being common knowledge that all players use the *same* game theory book. This is needed to justify the requirement that the (possibly mixed) strategies that the players will use are common knowledge. A rational[4] player will necessarily optimize given his beliefs. These two considerations are inconsistent unless a Nash equilbrium is employed.

Sometimes this story is embellished by a reference to negotiations which supposedly take place before the game is played. Usually, the negotiation is said to take place "in the bar the night before." At this meeting, the players jointly reason their way to an agreement on how to play tomorrow in the *absence* of any mechanism which allows them to make *binding* commitments on tommorrow's play. The players can then be regarded as arriving at the game venue equipped with a jointly authored book (perhaps typed overnight by the bartender) and *with good reason* to believe that the other players are carrying the *same* book.

The motivation for this embellishment is the feeling that the traditional defence of Nash equilibria is a little thin without some explanation of *how* it came to be common knowledge that the players are all using the same game theory book. Clearly some explanation is required but the "meeting in the bar the night before" will not suffice for the purposes of this book.[5] My own view is that if such a meeting really takes place then the opportunities for communication at the meeting should be modeled as formal moves in an enlarged game which should then be analyzed as a *contest* in accordance with the Nash program (section 2.1). In particular, one has to proceed *without* the device of a "meeting in the bar the night before."

There is, in any case, the more fundamental question of why we are to study rational agents at all. The traditional answer is that evolutionary processes (social and economic, as well as biological) will tend to remove behavior from the scene which is not well adapted to the environment. In so far as this answer is adequate, it "explains" not only the rationality of the agents but their treatment of each other as though they were all rational in the same sense. One advantage of this point of view is that agents are not seen as omniscient mathematical prodigies. Cyclists provide a suitable analogy. In keeping their balance, they do not solve complex systems of differential equations in their heads (or, at least, not consciously). Nor are they conscious of monitoring such factors as windspeed and the camber of the road. Nevertheless, they behave *as though* they were consciously gathering all necessary data and then computing an optimal response. This is ensured by a simple selection mechanism. Either a prospective bicycle rider learns to behave this way or he does not become a bicycle rider.

Such matters are discussed more expansively in later chapters. Here all that will be attempted is an illustration of how Nash equilibria can emerge from mechanical evolutionary processes.

Suppose that the game of example 2.3.5 is played repeatedly by two robots. Each robot is programmed to play as follows. If the game has already been played n times and the opposing robot chose its second pure strategy on m of these occasions, then the robot with which we are concerned begins by calculating the relatively frequencies $(n-m)/n$ and m/n with which its opponent has used its first and second pure strategies respectively. It then chooses the pure strategy[6] for its next play which *would be* optimal *if it were the case* that the opponent was going to choose his first pure strategy next time with probability $(n-m)/n$ and its second pure strategy with probability m/n. The robots also need a rule to get them started in the first place. For our purposes, it does not matter what this rule is. In fact, we can proceed by supposing that the robots operate according to some unknown choice rules for the first N plays and only adopt the choice rule given above when $n > N$.

Our choice rule is said to be *myopic* because it disregards relevant information. This means that the predictions the players make about each others' future play are not as good as they could be. A human replacement for one of our robots, for example, might eventually guess the choice rule for its opponent and then be able to predict its future behavior precisely.

The model, as it stands, is not complicated but its analysis requires some attention to detail. We therefore adopt the mathematician's trick of moving from discrete time to continuous time. Let $p(t)$ denote the relative frequency with which player I has chosen his second pure strategy up to time t and use $q(t)$ similarly for player II. If $(p(t), q(t))$ lies in the shaded region of figure 2.12(a) (reproduced from figure 2.7(b)), then player I will be choosing its second pure strategy and player II will be choosing its first pure strategy during the time interval $[t, t+\tau]$, provided $\tau > 0$ is sufficiently small. But then

$$p'(t) = \lim_{\tau \to 0+} \frac{p(t+\tau) - p(t)}{\tau} = \lim_{\tau \to 0+} \frac{1}{\tau}\left[\frac{\lambda t p(t) + \lambda \tau}{\lambda(t+\tau)} - p(t)\right] = \frac{1 - p(t)}{t}$$

$$q'(t) = \lim_{\tau \to 0+} \frac{q(t+\tau) - q(t)}{\tau} = \lim_{\tau \to 0+} \frac{1}{\tau}\left[\frac{\lambda t q(t)}{\lambda(t+\tau)} - q(t)\right] = -\frac{q(t)}{t}$$

where λt measures the number of games played during a time period of length t. We can solve the first differential equation by observing that it reduces to $(pt)' = p't + p = 1$. Hence $1 - p(t) = a/t$ where a is a constant of integration. Similarly, $q(t) = b/t$. Thus

$$\frac{q(t)}{1 - p(t)} = \frac{b}{a}$$

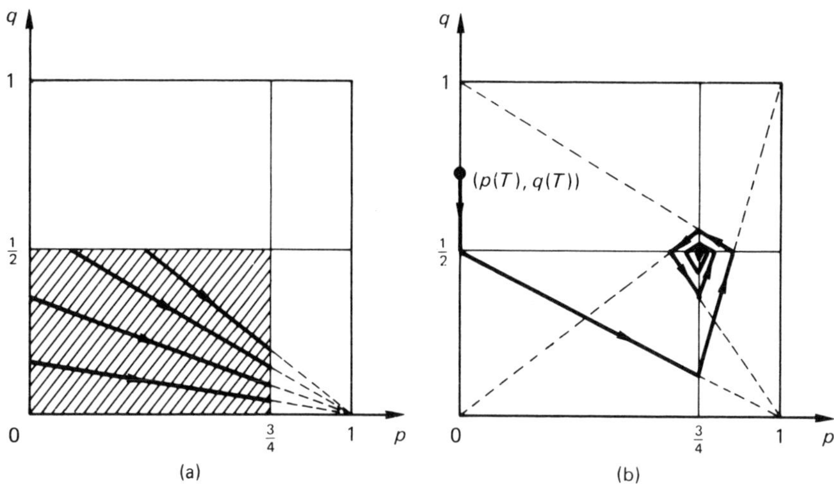

Figure 2.12

It follows that $(p(t), q(t))$ moves along one of the trajectories illustrated in figure 2.12(a) as long as it remains in the shaded region.

Figure 2.12(b) shows how the trajectories look in the four regions and traces how $(p(t), q(t))$ will evolve over time if it starts at the indicated point at time T. It will be obvious that the same cobweb pattern converging on the Nash equilibrium at $(1/2, 3/4)$ will appear wherever $(p(T), q(T))$ may be located.

A kitbitzer will see both robots sometimes playing one pure strategy and sometimes the other. If he calculates how frequently each pure strategy is played, he will find that, in the long run, the relative frequencies approach their Nash equilibrium values. In this sense each robot's limiting behavior is as rational as that of Von Neumann himself. But the semblance of rationality has been produced mindlessly by a *myopic tâtonnement*[7], i.e. a short-sighted (or mechanical) adjustment process.

Essentially the same mathematics can be used to tell a different evolutionary story with a similar punch line. Imagine two populations (A and B) of potential players of the same game as before (example 2.3.5). Over time, pairs of individuals are selected at random from the two populations to play the game. Player I comes from population A and player II comes from population B. Before entering the game-playing population, individuals briefly attend a gaming school where they are taught the "correct" strategy which, later in life, they never think to question. However, what is taught in school changes over time. At any time t, there will therefore be a probability $p(t)$ that a randomly chosen member of population A will choose pure strategy 2 if required to act as player I and a probability $q(t)$ that a randomly chosen member of population B will choose pure strategy 2 if required to act as player II. The school system is relatively enlightened in that the strategy it designates as "correct" is always optimal *at the time the instruction is given*, i.e. for the current values of $p(t)$ and $q(t)$. To obtain simple differential equations, we assume that the number N in each population is very large and that the probability of any given member of the population being retired from the game-playing scene in an interval of length t is λt. The numbers in both populations are held steady by recruiting school-leavers in just the right numbers to offset retirements. Thus $N\lambda\tau$ school-leavers enter each population between time t and $t+\tau$. If $(p(t), q(t))$ lies in the shaded region of figure 2.12(a), then

$$p'(t) = \lim_{\tau \to 0+} \frac{1}{N\tau}[Np(t)(1-\lambda\tau) + N\lambda\tau - Np(t)] = -\lambda p(t) + \lambda$$

$$q'(t) = \lim_{\tau \to 0+} \frac{1}{N\tau}[Nq(t)(1-\lambda t) - Nq(t)] = -\lambda q(t)$$

These differential equations are linear with solutions $1-p(t) = ae^{-\lambda t}$ and $q(t) = be^{-\lambda t}$. Thus, in the shaded region of figure 2.12(a),

$$\frac{q(t)}{1-p(t)} = \frac{b}{a}$$

just as in the robot example. As before, $(p(t), q(t))$ converges over time along the cobweb path shown in Figure 2.12(b) to the Nash equilibrium (1/2, 3/4).

These stories about myopic *tâtonnement* generating Nash equilibria are only stories. There is no difficulty in constructing a myopic *tâtonnement* which does not converge to anything. Simply let each robot in the first story optimize on the assumption that its opponent will play precisely as it did in the previous game. Play will then cycle indefinitely as indicated in figure 2.13. One might say that the environment is changing too fast for the players' behavior to have a change of adapting to it. The stories therefore do not prove anything about the convergence of *tâtonnement* processes in general. Nor are they intended as acceptable representations of any *tâtonnement* process that

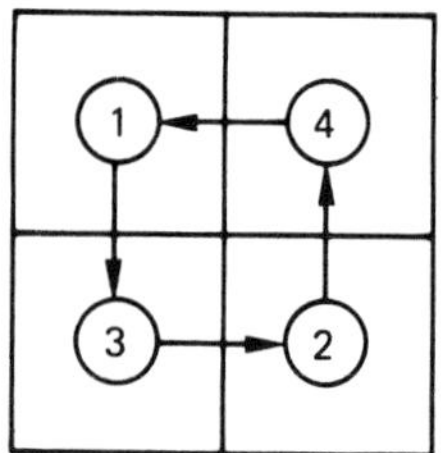

Figure 2.13

actually exists. They are parables. Their purpose is to illustrate a philosophical point, not to make a statement about the way the world is.

This brings us to the interpretation of mixed strategies. It is sometimes thought to be paradoxical that rational considerations should lead players to make decisions at random. The traditional explanation is that any line of thought that a rational player may pursue can be duplicated by a rational opponent.[8] Hence the only way a rational player can prevent his choice of pure strategy being accurately predicted by a rational opponent is by resorting to randomization. Perhaps a paradox is seen here by some because the use of randomization in real life is often disguised, the randomizing device not being recognized as such. For example, it is often important that the timing of governmental decisions on economic matters not be predicted by speculators. This can sometimes be achieved by delegating the decision to a committee of "experts" whose role is to determine precisely the right time to act. Since the behavior of such committees is largely unpredictable, they serve as effective randomizing devices.

Tâtonnement models allowed a second interpretation of mixed strategies. Only in very exceptional cases will myopic decision rules lead to indifferences between the available pure strategies. In each game, both players will therefore nearly always choose a pure strategy, but, in different games, different pure strategies will sometimes be chosen. A mixed strategy can then be seen as a statement about how many times *on average* a pure strategy is used. In the example of repeated play between two robots, the average is taken over *time*. After long enough, player I will have used its second pure strategy about 3/4 of the time. In the example in which players were sampled from large populations, the average is taken over a population. After long enough, about 3/4 of population A will be playing their second pure strategy when acting as player I.

Some concluding remarks are necessary to put what has been said in this section into perspective.

One way of viewing the manner in which we have sought to justify Nash equilibria is that two distinct interpretations are possible. In Binmore (1987 – this volume, ch. 5), these are referred to as the *eductive*[9] interpretation and the *evolutive* interpretation. In the eductive interpretation, the players are thought of as educated well-informed individuals equipped with an authoritative game theory book whose contents have been deduced from first principles by logico-rational means. In the evolutive interpretation, the players are thought of as stimulus–response machines whose behavior has become adapted to their environment as a result of the removal of unadapted behavior via the operation of a trial-and-error adjustment process. It is important, however, to see that the distinction between these two interpretations is quantitative rather than qualitative.

In the first place, it is not necessary that evolutive processes be confined to stimulus–response machines whose internal complexity is comparable with that of a chocolate-dispensing machine. It is true that explicit mathematical models of adaptive processes, like those given in this chapter, usually restrict themselves to very simple (myopic) machines and it is interesting that knee-jerk behavior can adapt so well if the circumstances are favorable. (Any book on evolutionary biology provides hosts of fascinating examples.) But it is not true that only simple machines are worth studying. The reason that the players are seldom modeled as computers with complex programs is that the consequent mathematical analysis is usually too difficult for the tools we currently have to hand. Nevertheless, a long-range target for the game theorist is the study of the *evolution of learning rules*. In this context, one thinks of the programs which operate the computers as competing for survival rather than the computers themselves. Where the computer is serving as a paradigm for one of the lower animals, the program has to be thought of as genetically determined and the competition as biological. Among the higher animals, including *homo sapiens*, the competition will also be social and economic – or, to put the same point more dramatically, Nature, red in tooth and claw, also operates in the world of ideas.

Once one has come this far, it becomes possible to ask whether rationality itself has any genuine meaning except in an evolutionary setting. (And, for the cognoscenti, we do *not* mean "Bayesian rationality." Here, and always, we intend rational in its common-or-garden dictionary sense. In particular, to be rational does not exclude the use of the scientific method.) One might think of rationality principles as second-level learning rules that sort and filter first-level learning rules. As such, they too will compete on the battleground of ideas. Even if one prefers to hang on to the notion of rationality as a Platonic ideal, its interest for real-world behavior lies largely in the extent to which it can serve as a paradigm for the result of competition for survival among an ever-changing population of rival rules of thumb.

In summary, I do not see the evolutive and eductive interpretations as distinct and separate. An evolutive scenario involving knee-jerk stimulus–response machines and an eductive scenario involving omniscient mathematical wizards are the two polar extremes on a continuous scale. In due course, it is to be hoped that current research into theories of "bounded rationality" will tell us more about the intermediate points on this scale.

2.6 Security levels

Consider an n-person game with strategy spaces $S_1, S_2, \ldots, S_n$. We shall write $S = S_1$ and $T = S_2 \times S_3 \times \ldots, \times S_n$. If the payoff functions are bounded, we define the first player's security level v_1 by

$$v_1 = \max_{s \in S} \min_{t \in T} \phi_1(s, t)$$

Security levels for the other players are defined similarly. The maximum and minimum values in the definition of a security level always exist when the strategy spaces are compact and the payoff functions are continuous (as, for example, in the case of finite games). In more complicated situations, it is sometimes necessary to write "sup" and "inf" instead of "max" and "min."

Suppose that

$$m(s) = \min_{t \in T} \phi_1(s,t)$$

and that the minimum is achieved at $\mu(s) \in T$. Then the choice of s by the first player *guarantees* an expected payoff to him of at least $m(s)$ because $m(s) \leqslant \phi_1(s, t)$ for all t that might be chosen by the other players. On the other hand, player I cannot *guarantee* getting more than $m(s)$ because it is open to the other players to choose $\mu(s)$ and $m(s) = \phi_1(s, \mu(s))$. Thus, $m(s)$ is the largest payoff that player I can count on by the play of s. It follows that

$$v_1 = \max_{s \in S} \phi_1(s, \mu(s)) = \max_{s \in S} \min_{t \in T} \phi_1(s, t)$$

is the largest payoff to player I which he can *guarantee* if he is free to choose his strategy s. We shall refer to the strategy $\hat{s}$ which guarantees a player's security level as a *security strategy*. Thus

$$v_1 = \phi_1(\hat{s}, \mu(\hat{s})) = \max_{s \in S} \min_{t \in T} \phi_1(s, t)$$

Because the payoffs in the above analysis are Von Neumann and Morgenstern utilities, we are only claiming that a security level can be guaranteed as an expectation. Thus $\hat{s}$ guarantees an *expected* payoff of at least v_1. This is important because it often happens that a security strategy turns out to be mixed. This is true in example 2.3.5. As we shall see, player I has a unique security strategy requiring him to play his second pure strategy with probability 3/4.

Two properties of security strategies, analogous to properties (1) and (2) for Nash equilibrium strategies given in section 2.4, deserve mention. A security strategy never assigns positive probability to a strongly dominated pure strategy; and a player always has at least one security strategy which assigns zero probability to a weakly dominated strategy. There is a further important property which also has a parallel for Nash equilibrium strategies. This says that a security strategy $\hat{s}$ guarantees player I's security level v_1 against all mixed strategies if and only if it guarantees v_1 against all pure strategies. The reason is that, although the other players may use random devices in choosing their pure strategies, nevertheless a pure strategy combination will eventually be played. If $\hat{s}$ guarantees v_1 against any such pure strategy combination, then it follows that v_1 is always guaranteed. (Similarly, a mixed strategy is an optimal response to the strategy choices made by the other players if and only if each of the pure strategies to which it attaches positive probability is an optimal response.)

Example 2.6.1

We now compute security strategies for the game of example 2.3.5 using the last of the properties for security strategies given above. If player I chooses his second pure strategy will probability p and player II chooses her first pure strategy for certain, then player I gets a payoff of

$$E_1(p) = 1(1-p) + 3p = 1 + 2\text{p}$$

If player II uses his second pure strategy for certain, then player II gets a payoff of

$$E_2(p) = 4(1-p) + 2p = 4 - 2p$$

In calculating what is guaranteed for player I, we need only consider that happens when player II uses pure strategies. Thus the choice of p by player I guarantees $\min\{E_1(p), E_2(p)\}$. We require the value $\hat{p}$ at which this is maximized, i.e.

$$v_1 = \min\{E_1(\hat{p}), E_2(\hat{p})\} = \max_p \min\{E_1(p), E_2(p)\}$$

Figure 2.14(a) makes it clear that $\hat{p}$ is the solution of the equation

$$1 + 2p = 4 - 2p$$

Hence $\hat{p} = 3/4$ and $v_1 = 5/2$. Figure 2.14(b) indicates that the corresponding value $\hat{q}$ for player II is the solution of

$$-(1 + 3q) = -(3 - q)$$

and so $\hat{q} = 1/2$ and $v_2 = -5/2$. (Recall that the matrix entries are negative payoffs for player II.)

It is no accident that the security strategies in this game constitute a Nash equilibrium (example 2.3.5). The next section establishes that this is a general result for a zero-sum matrix game. For bimatrix games, the connection between Nash equilibria and security strategies is less close. It is easy to

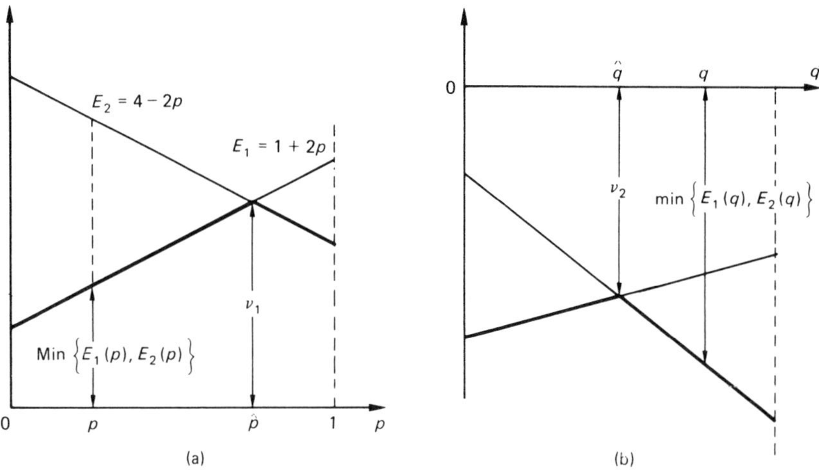

Figure 2.14

prove that it is always true that

$$v_1 = \max_{s \in S} \min_{t \in T} \phi_1(s, t) \leqslant \min_{t \in T} \max_{s \in S} \phi_1(s, t)$$

Moreover, if $(\tilde{s}, \tilde{t})$ is a Nash equilibrium of the game, then it is equally easy to prove that

$$\min_{t \in T} \max_{s \in S} \phi_1(s, t) \leqslant \phi_1(\tilde{s}, \tilde{t})$$

with similar inequalities for player II. It follows that a Nash equilibrium gives each player at least his security level. Beyond this, there is not a great deal that can usefully be said.

2.7 Von Neumann's minimax theorem

So far the treatment of zero-sum games has been unconventional in that no mention has been made of Von Neumann's famous minimax theorem.[10] This will be deduced from Nash's theorem with the aid of the following lemma. (An alternative proof can be based on the argument given in example 2.3.7.)

SADDLE-POINT LEMMA

In a two-person zero-sum game, the following conditions are equivalent.

(i) $(\tilde{s}, \tilde{t})$ is a Nash equilibrium

(ii) $(\tilde{s}, \tilde{t})$ is a saddle point for ϕ_1, i.e. for all $s \in S$ and all $t \in T$,

$$\phi_1(s, \tilde{t}) \leqslant \phi_1(\tilde{s}, \tilde{t}) \leqslant \phi_1(\tilde{s}, t)$$

(iii) $\tilde{s}$ is a security strategy for player I, $\tilde{t}$ is a security strategy for player II, and $v_1 = \phi_1(\tilde{s}, \tilde{t}) = -v_2$.

(Note that the lemma remains true if the game is strictly competitive rather than zero sum, except that v_2 is not necessarily equal to $-\phi_1(\tilde{s}, \tilde{t})$. Instead, we must write $v_2 = \phi_2(\tilde{s}, \tilde{t})$.)

Proof. To see that (i) says the same as (ii), it is only necessary to observe that $\phi_2(\tilde{s}, \tilde{t}) \geqslant \phi_2(\tilde{s}, t)$ if and only if $\phi_1(\tilde{s}, \tilde{t}) \leqslant \phi_1(\tilde{s}, t)$.

To prove that (ii) implies (iii), observe that the inequality $\phi_1(\tilde{s}, \tilde{t}) \leqslant \phi_1(\tilde{s}, t)$ means that the strategy $\tilde{s}$ guarantees at least $\phi_1(\tilde{s}, \tilde{t})$ for player I. Thus $v_1 \geqslant \phi_2(\tilde{s}, \tilde{t})$. The inequality $\phi_2(s, \tilde{t}) \leqslant \phi_1(\tilde{s}, \tilde{t})$ means that player II can guarantee that player I gets at most $\phi_1(\tilde{s}, \tilde{t})$ by playing $\tilde{t}$. Thus $v_1 \leqslant \phi_1(\tilde{s}, \tilde{t})$. It follows that $\tilde{s}$ is a security strategy for player I and $v_1 = \phi_1(\tilde{s}, \tilde{t})$. Similarly, $\tilde{t}$ is a security strategy for player II and $v_2 = \phi_2(\tilde{s}, \tilde{t}) = -\phi_1(\tilde{s}, \tilde{t})$. In particular, $v_1 = \phi_1(\tilde{s}, \tilde{t}) = -v_2$.

To prove that (iii) implies (i), note that, if $(\tilde{s}, \tilde{t})$ is *not* a Nash equilibrium, then at least one player can benefit from a deviation (provided that the opponent does not deviate). Suppose that player I benefits by switching from $\tilde{s}$ to s. Then player I gets *more* than his security level v_1. But the game is zero-sum and $v_1 + v_2 = 0$. Hence player II gets *less* than her security level v_2 after the deviation. But $\tilde{t}$ guarantees her security level *whatever* player I does. This is a contradiction.

VON NEUMANN'S THEOREM

In a zero-sum matrix game in which S and T are the sets of all mixed strategies,

$$\max_{s \in S} \min_{t \in T} \phi_1(s, t) = \min_{t \in T} \max_{s \in S} \phi_1(s, t)$$

Proof. The theorem simply states that $v_1 = -v_2$ and hence follows, via the saddle-point lemma, from the version of Nash's theorem which says that matrix games always have at least one Nash equilibrium.

Von Neumann and Morgenstern argued directly for the use of security strategies in zero-sum matrix games, or, to express the same point more conventionally, they advocated the use of a "maximin decision rule." In zero-sum matrix games, the rule not only has the advantage that it generates a Nash equilibrium, as is appropriate for play between rational players; it also insures players against irrational play by the opponent. Since the equilibrium strategies are secure, a player gets an expected payoff at least as

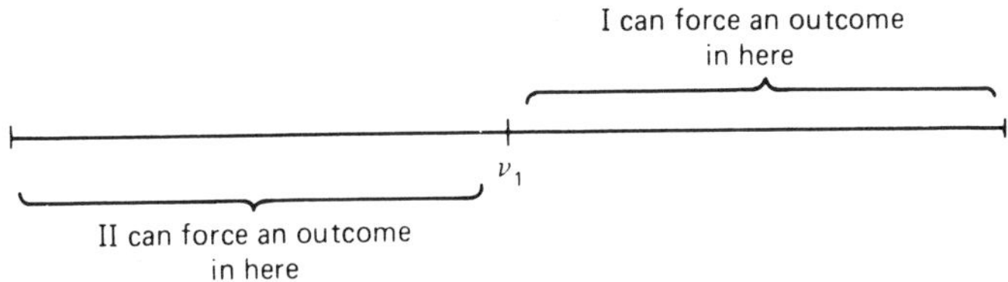

Figure 2.15

large as his or her equilibrium payoff whatever the opponent does. The situation is illustrated by figure 2.15 in which the line segment represents the set of payoffs to player I which could result from the play of the zero-sum game in question.

Let us consider this insurance argument in favor of the maximin principle a little more closely. It is certainly true that my opponent in any game will necessarily have gone badly astray in predicting my behavior if he ends up with less than his security level. A well-informed rational player will presumably not go astray like this in his reasoning. Hence, if my opponent in a zero-sum matrix game is well informed and rational, then I cannot get more than my security level. It follows that the play of one of my security strategies must be optimal for me. Similarly for my opponent.

Note that this argument does not provide me with any genuinely good reason to play a security strategy against an opponent whom I perceive to be irrational. Against such an opponent, I would wish to *deviate* from my security strategy in order to exploit his anticipated bad play. For example, if player II's security strategy is completely mixed (example 2.3.6), then player I's security strategy always generates the *same* expected payoff to player I whether player II sticks to her security strategy or not.[11] Maximin play, under such circumstances, therefore rules out completely any prospect of cashing in on perceived weaknesses in the opponent.

We therefore have little cause for advocating maximin play to rational players in zero-sum matrix games beyond those already given in support of Nash equilibria in section 2.5, and no cause whatsoever to advocate maximin play to rational players in other games. Indeed, as later examples will show, the maximin principle has little to recommend it as a general decision-making principle in game theory.

Finally, it should be noted that, although the insurance argument is not very persuasive when it comes to giving advice to rational players, it does not follow that it has no relevance when equilibrium is to be achieved, if at all, through the operation of a myopic adaptive process. *Whatever* stage the adaptive process has reached, at least one of the players in a zero-sum matrix game will always be no worse off after switching to a security strategy and usually will be better off.

2.8 The value of a game

The *value* of an *n*-person game is the vector $\mathbf{v} = (v_1, v_2, \ldots, v_n)$ of expected payoffs that the players will receive if they play as their solution strategies recommend. Such a statement presupposes that solution strategies of the game exist and we know what they are. It is not enough, even in the case of a static contest, to argue that the solution strategies must constitute a Nash equilibrium. In general, a game may have many Nash equilibria. The question then arises: which of these equilibria is to be selected as the solution? This is the *equilibrium selection problem* about which nothing very much will be said here. At this point, the only comment which ought to be made is that the answer might be: none at all. In so far as an argument has been offered in favor of Nash equilibria, it asserts only that, *if* a solution exists, then it must be a Nash equilibrium.

The selection problem becomes vacuous in the case of static two-person strictly competitive games. To explain why, we introduce two definitions.

1 *Equivalent.* Two Nash equilibria (s_1, s_2) and (t_1, t_2) are said to be equivalent if each player receives the same payoff at (s_1, s_2) as he or she receives at (t_1, t_2), i.e.

$$\phi_1(s_1, s_2) = \phi_1(t_1, t_2)$$

$$\phi_2(s_1, s_2) = \phi_2(t_1, t_2)$$

2 *Interchangeable.* Two Nash equilibria (s_1, s_2) and (t_1, t_2) are said to be interchangeable if and only if the strategy pairs (s_1, t_2) and (t_1, s_2) are also Nash equilibria.

The saddle-point lemma (section 2.7) ensures that, for a static two-person strictly competitive game, the set of all Nash equilibria is the set of all pairs of security strategies for the game (provided that at least one Nash equilibrium actually exists). It follows immediately that all Nash equilibria in such games are *interchangeable.* Moreover, the saddle-point lemma also asserts that all Nash equilibria assign each player precisely his or her security level. Thus all Nash equilibria are *equivalent.*

The point of introducing these definitions is that, if all Nash equilibria for a game are equivalent and interchangeable, then it *does not matter* if two game theorists disagree about which of them is to be selected as the solution of the game. If player I listens to Von Neumann and is persuaded that (s_1, s_2) is the solution, then he will choose strategy s_1. If player II simultaneously listens to Morgenstern and is persuaded that (t_1, t_2) is the solution, then she will choose strategy t_2. But (s_1, t_2) remains a Nash equilibrium and both players get precisely the payoffs they were promised.

For these reasons, a static two-person strictly competitive game is regarded as having an unambiguous value, namely (v_1, v_2). For zero-sum games, it is usual to quote only the payoff for player I. The value of such a game is then simply said to be v_1.

2.9 Trembling-hand equilibria

Various so-called "refinements" of the Nash equilibrium idea have been proposed (for a survey, see Van Damme, 1983). Mostly, these are based on the principle that the strategies the players use should be stable in the presence of small perturbations in the world about them. This certainly makes good sense when the equilibrium is interpreted as the end-product of a real-time myopic *tâtonnement*. Small random fluctuations of one kind or another are inevitable in any evolutionary process. Either an equilibrium will be stable in respect of the fluctuations to which it is subject, or else the system will not remain in equilibrium. The idea that an equilibrium may be "stable" or "unstable" is no different in essence from the same idea in physics. A soccer ball resting at the apex of a hill is in unstable equilibrium because a slight puff of wind will be enough to set it rolling. The same ball at the bottom of a pit is in stable equilibrium because, after any small disturbance, it will roll back to its resting place.

The difficulty with applying this idea in game theory lies in deciding what type of perturbation ought to be investigated. Obviously, this will depend on the nature of the dynamic process by means of which equilibrium is achieved. But the abstract definition of a static game provides no information at all about *how* equilibrium is achieved. To try to deduce the "correct" perturbations from the mathematical formulation of the game would therefore be like trying to deduce the "correct" solution of a quadratic equation from a study of the values of its coefficients. If perturbations are to be studied in the context of a myopic *tâtonnement*, then the type of perturbations chosen for study will necessary reflect some judgment about the nature of the *tâtonnement* process. Such judgments import further information into the problem over and beyond that provided by the formal definition of a static game.

One type of perturbation which it makes sense to study, in certain contexts, consists of small random errors which intervene between a player's decision to act and the act itself. A player's hand is thought of as "trembling" as it reaches for the strategy buttons and sometimes pressing the wrong button by mistake. Selten (1975) studied the limits of the Nash equilibria of such perturbed games as the perturbations were allowed to become very small. He found that these limits are identical with the "trembling-hand equilibria" defined below.

A *trembling-hand equilibrium* for a two-person matrix game is a pair of strategies $(\tilde{\mathbf{p}}, \tilde{\mathbf{q}})$ such that $\tilde{\mathbf{p}}$ is an optimal response to *each* term of at least one sequence $\{\mathbf{q}_k\}$ of completely mixed strategies such that $\mathbf{q}_k \to \tilde{\mathbf{q}}$ as $k \to \infty$ and $\tilde{\mathbf{q}}$ is simultaneously an optimal response to *each* term of at least one sequence $\{\mathbf{p}_k\}$ of completely mixed strategies such that $\mathbf{p}_k \to \tilde{\mathbf{p}}$ as $k \to \infty$. Recall that a completely mixed strategy attaches *positive* probability to *every* pure strategy. To say that $\tilde{\mathbf{p}}$ is an optimal response to each term of $\{\mathbf{q}_k\}$ means that

$$\phi_1(\tilde{\mathbf{p}}, \mathbf{q}_k) \geqslant \phi_1(\mathbf{p}, \mathbf{q}_k) \qquad (k = 1, 2, \ldots)$$

for all mixed strategies $\mathbf{p}$. Since the payoff function ϕ_1 for a matrix game is linear (and hence continuous) in $\mathbf{q}$ for each fixed $\mathbf{p}$, it follows that

$$\phi_1(\tilde{\mathbf{p}}, \tilde{\mathbf{q}}) \geqslant \phi_1(\mathbf{p}, \tilde{\mathbf{q}})$$

for all mixed strategies $\mathbf{p}$. Thus a trembling-hand equilibrium is necessarily a Nash equilibrium.

But the converse does not hold. For a trembling-hand equilibrium, $\tilde{\mathbf{p}}$ is not only an optimal response to $\tilde{\mathbf{q}}$, it is an optimal response to a whole sequence $\{\mathbf{q}_k\}$ of perturbations of $\tilde{\mathbf{q}}$. Thus the property that $\tilde{\mathbf{p}}$ is an optimal response is *stable* given these perturbations. Similarly, the property that $\tilde{\mathbf{q}}$ is an optimal response is stable in the presence of a sequence of perturbations of $\tilde{\mathbf{p}}$.

We know that a Nash equilibrium strategy always attaches zero probability to *strongly* dominated strategies. A trembling-hand equilibrium always attaches zero probability to *weakly* dominated strategies also. Example 2.3.3 is instructive on this point. Our analysis of some pages back shows that the value of this zero-sum matrix game is 3. Player I has a unique security strategy consisting of his second pure strategy. Any mixed strategy of the form $(1-q, 0, q)^{\mathrm{T}}$ is a security strategy for player II.

It is never optimal for player II to use her third pure strategy in this game if there is a positive probability that player I will choose his first pure strategy. It follows that the game has a *unique* trembling-hand equilibrium, namely the pure strategy profile (row 2, column 1).

Can we now claim that the "trembling-hand refinement" of Nash equilibrium has provided us with unique solution strategies for the game (as well as a unique value)? Following Selten (1975), trembling-hand equilibria are usually called *perfect equilibria* precisely because it is believed that those Nash equilibria which are not trembling-hand equilibria are "imperfect" and hence should be rejected as solution candidates. To quote Selten (1975, p. 35):

> There cannot be any mistakes if the players are absolutely rational. Nevertheless, a satisfactory interpretation of equilibrium points in extensive games seems to require that the possibility of mistakes is not completely excluded. This can be achieved by a point of view which looks at complete rationality as a limiting case of incomplete rationality.

This view seems unassailable in so far as it asserts that it is necessary to contemplate perturbations of whatever we decide to use as our model of a rational player. But there are many other ways of perturbing a rational player than by jogging his elbow (and, for a game like chess, many of these will provide a more plausible explanation of a deviation from an anticipated line of play than that the opponent moved a piece he did not mean to move by mistake). However, as the quote indicates, this problem is more serious in extensive (or dynamic) games than it is for the static game of figure 2.16(a).

But, even in such a very simple game, matters are not as cut and dried as some authors would have us believe. Figure 2.16(b) shows a trajectory diagram (analogous to those of figure 2.10) for the *tâtonnement* models of section 2.5 but applied to our current example with the strongly dominated strategy (column 2) deleted. In section 2.5, the question of what players do when they are indifferent between two courses of action was a minor distraction and so we relegated to a note the observation that they then choose at random between the two alternatives. Here this proviso actually matters in that, if a point $(1, y)$ in figure 2.16(b) is ever reached, then the players will drift to the final outcome $(1, 1/2)$ and *not* to the trembling-hand equilibrium $(1, 0)$. Bear in mind, before rejecting the random choice mechanism we have described for distinguishing between equally attractive actions, that our story is not so very different from that use by biologists

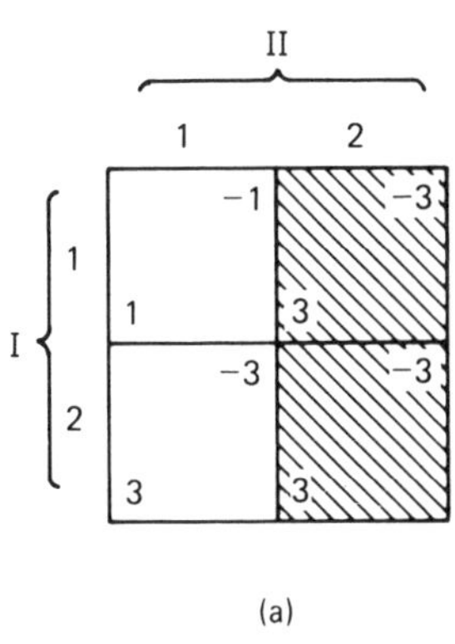

(a)

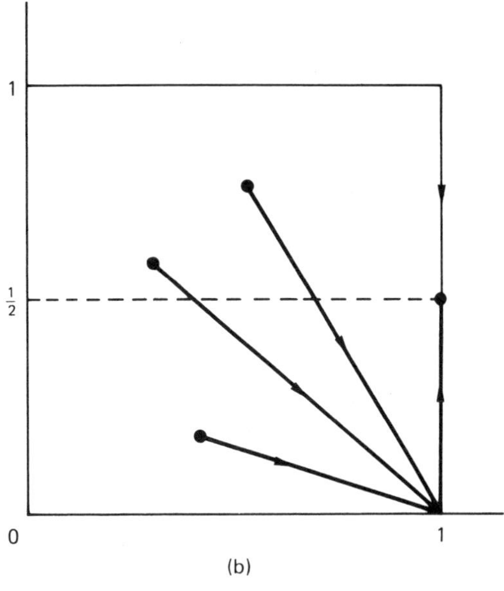

(b)

Figure 2.16

when discussing the phenomenon of "genetic drift" in respect of traits which are not under adaptive pressure.

In summary, I deny that mathematical stability definitions can legitimately be used as automatic filters to refine the set of Nash equilibria without generating implicit restrictions on the class of models for which the game is a useful analytic tool. This is not to say that refinements should not be used, but only that great care must be taken to tailor the refinement used to the situation in which it is to be applied. Nor is it to say that weakly dominated strategies should never be discarded: only that, when this is done, note should be taken that something extra is being assumed.

NOTES

1 Although it should be noted that recent work of Kohlberg and Mertens (1986) revives the Von Neumann and Morgenstern view but in a rather more sophisticated manner than that in which it was originally advanced.

2 One might argue that both players could condition their behavior on an event like the number of times the letter "e" appears in today's New York Times or, to quote an example popular in economics, on the current level of sunspot activity. The players' choices would then be random but *correlated*. But how do the players fix on a random event to observe jointly if no pre-play communication, either explicit or implicit, is allowed?

3 This will be true if and only if $-\phi_1 = A\phi_2 + B$, where $A > 0$ and B are constants.

4 The "Bayesian" view is that this is what "rational" *means*. Section 2.2 contains an explanation of why I believe such a view to be inadequate.

5 Or, indeed, for many of the economic models for which Nash equilibrium is used. Where it is *known* that no pre-play negotiation took place, it seems futile to use a concept justified by assuming such pre-play negotiations. And where pre-play negotiations do take place, the appropriate concept would seem to be Aumann's "correlated equilibrium" rather than Nash equilibrium.

6 If both pure strategies happen to be optimal, we may allow the robot to choose randomly between them.

7 Moulin (1986) prefers the even more gallic "myopic *tâtonnement à la Cournot*." Note that I do not use *tâtonner* in any technical sense, but with its dictionary meaning of "to grope," "to fumble," "to feel one's way."

8 This raises a problem of self-reference which lives at the heart of traditional thinking. This is not the place to raise such issues but see, for example, Binmore (1987 – this volume, ch. 5).

9 Every humour hath its proper eductive cathartic (*Oxford English Dictionary*).

10 Which would be more aptly described as the "maximin theorem."

11 Let $(\tilde{\mathbf{p}}, \tilde{\mathbf{q}})$ be a Nash equilibrium for a bimatrix game with matrices A and B. Any pure strategy assigned positive probability by $\tilde{\mathbf{q}}$ is an optimal response to $\tilde{\mathbf{p}}$. Thus, if $\tilde{\mathbf{q}}$ is completely mixed, all pure strategies for player II are optimal responses to $\tilde{\mathbf{p}}$ and hence any mixture $\mathbf{q}$ of these is an optimal response to $\tilde{\mathbf{p}}$. In

a zero-sum game, this means that the play of $\mathbf{q}$ against $\tilde{\mathbf{p}}$ yields a payoff of v_2 to player II and hence of v_1 to player I.

REFERENCES

Aumann, R. 1976: "Agreeing to disagree," *Annals of Statistics* 4, 1236–9.

Binmore, K. 1987: "Modelling rational players I," *Economics and Philosophy* 3, 179–214.

Kohlberg, E., and Mertens, J.-F. 1986: "On the strategic stability of equilibria," *Econometrica* 50, 863–94.

Luce, R., and Raiffa, H. 1957: *Games and Decisions.* New York: Wiley.

Moulin, H. 1986: *Game Theory for the Social Sciences* (2nd edn). New York: New York University Press.

Nash, J. 1951: "Non-cooperative games," *Annals of Mathematics* 54, 286–95.

Owen, G. 1982: *Game Theory* (2nd edn). New York: Academic Press.

Selten, R. 1975: "Reexamination of the perfectness concept for equilibrium in extensive games," *International Journal of Game Theory* 4, 22–25.

Van Damme, E. 1983: *Refinements of the Nash Equilibrium Concept* (Lecture Notes in Economics and Mathematical Systems). Berlin: Springer.

Von Neumann, J., and Morgenstern, O. 1944: *The Theory of Games and Economic Behavior.* Princeton, N.J.: Princeton University Press.

3 Information

"I believe, John," pursued Bella, "that you believe that I believe ..."
"My dear child," cried her husband gaily, "what a quantity of believing."

Charles Dickens, *Our Mutual Friend*

3.1 Incomplete information

If players are to compute their Nash equilibrium strategies for a static contest, then they need to know what the game *is*. This is not a great difficulty when Nash equilibria are being interpreted in terms of a myopic *tâtonnement*. In this context, players' behavior adopts to the environment over time and, although the players may not "know" what game they are playing, nevertheless, in the long run, they act *as though* they knew and were optimizing on the basis of this knowledge.

However, for a story based on conscious decisions made by rational agents, the difficulty is a pressing one. So far we have proceeded on the implicit assumption that information is *complete* in the sense of Von Neumann and Morgenstern (1944). In the case of a static game, this means that the strategy spaces $S_1, S_2, \ldots, S_n$ and the payoff functions $\phi_i\colon S_1 \times S_2 \times \ldots \times S_n \to \mathbb{R}$ $(i = 1, 2, \ldots, n)$ are *common knowledge* (Aumann, 1976). Everybody knows this data, every knows that everybody knows, everybody knows that everybody knows that everybody knows; and so on. The assumption is harmless enough in respect of the strategy spaces[1] $S_1, S_2, \ldots, S_n$. But it is not harmless to assume that one player knows another's preferences with precision and even less harmless to assume that the second player knows that the first player knows.

If this assumption is relaxed, then the players will be uncertain about the preferences of their opponents. We can model such uncertainties by allowing players to attach probabilities to the various utility functions they believe might possibly be correct for their opponents. But this is not enough. To analyze the game, a player needs to be equipped, not only with beliefs about the preference structure of the game, but also with beliefs about the beliefs of the other players. Otherwise the player will have no basis for making predictions about what the other players will do. But, if player I is conditioning his behavior on beliefs about beliefs, then the other players will need to have beliefs about his beliefs about beliefs if they are to be able to anticipate his behavior.

We seem to be led into an infinite regress. However, Harsanyi's (1967–8) theory of incomplete information[2] cuts the Gordian knot. We describe his idea (in a form he attributes to Selten) for the case of a two-player game.

Harsanyi pastulates the existence of a notional "chance move" which operates before the game is played. This selects a pair of *actors* to fill the *roles* I and II. Once the actors have been selected, they play the game in the usual way. The actors for each role are drawn from two predetermined populations, each actor being defined by a specification of his or her

1 *tastes*, summarized by a utility function defined on the outcomes of the game, and
2 *beliefs*, summarized by a probability distribution over the possible outcomes of the initial chance move.

All this data is assumed to be common knowledge. Common knowledge about individuals is therefore replaced by common knowledge about populations.

The game is now analyzed as a game of *complete* information in which the players are *not* identified with the *roles* I and II but with the *actors* who may fill these roles. A choice of strategy by an actor is interpreted as meaning that this is the strategy he or she will play *if* chosen to play the game by the initial chance move. (If an actor is not chosen to play the game, we may assume that he or she receives a payoff of zero regardless of what happens otherwise.) As always, we take it that rationality requires that actors seek to maximize utility given their beliefs.

In summary, for a two-player game, we begin with a pair of strategy spaces S_1 and S_2. This pair constitutes what is sometimes called a "game form." Pursuing our theatrical analogy, one might think of (S_1,S_2) as a *script* for a game with two *roles* I and II. Each role is then filled by one of a number of *actors* who are characterized by their *tastes* and *beliefs*. The pool of available actors is then treated as the set of "players" in an enlarged game in which information is *complete*. Harsanyi's theory therefore replaces a "game of incomplete information" by a "game of complete information." The cost is a potentially very large increase in the number of players and an increase in the complexity of the game to be analyzed. However, the introduction of the chance move does not destroy the static nature of the game. The actors all choose their strategies simultaneously *before* the change move operates. Assuming the game is to be analyzed as a contest, it is therefore still a Nash equilibrium that we have to look for. If the strategy profile chosen by the actors constitutes a Nash equilibrium, then no actor who is chosen with positive probability will have cause to deviate from his or her choice once the actor learns that he or she has been selected to play.

We postpone comments on terminology and related questions until after the following instructive example of Harsanyi (1967–8).

Example 3.1.1

In this example, the population from which an actor is drawn to fill role I is the set $\{a, b\}$. The population from which an actor is drawn to fill role II is the set $\{c, d\}$. Once actors have been chosen, a zero-sum 2×2 matrix game is played. Since the actors have different preferences, the payoffs in this game depend on which actors have been chosen to fill the two roles. We therefore have four possible payoff matrices, A_{ac}, A_{ad}, A_{bc}, and A_{bd}, to consider. These are indicated in figure 3.1. As always in two-person zero-sum games, only the payoffs for the actor in role I are entered. The actor in role II gets minus this amount.

The shaded cells show what the solutions of the four zero-sum 2×2 matrix games *would be* if the identities of the actors *were to become* common knowledge after their selection. Dominance arguments suffice to check this assertion. But it will not be common knowledge who has been selected.

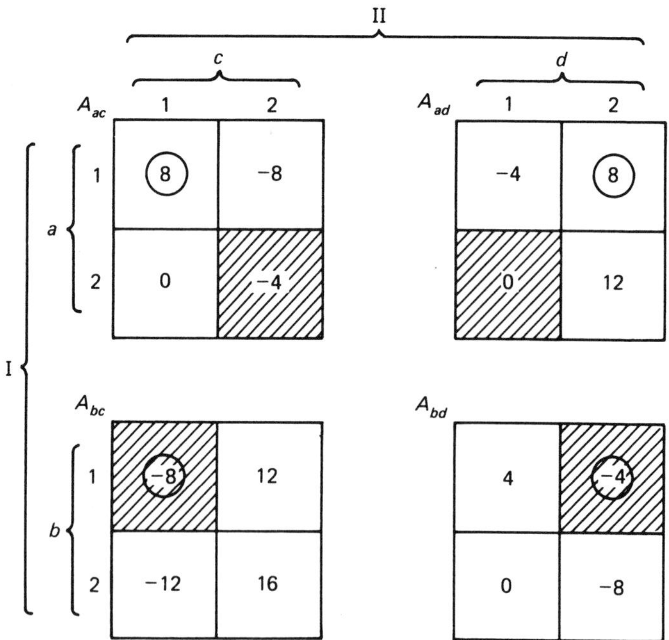

Figure 3.1

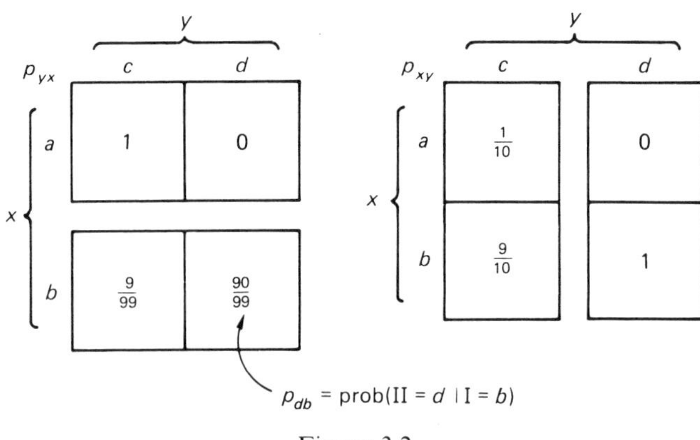

Figure 3.2

Harsanyi's theory requires instead that the description of an actor x should incorporate the beliefs that x will have about the identity of the actor y chosen to fill the other role, after x learns that he or she has been called upon to play. We use the notation $p_{yx} = \text{prob}(\text{II} = y | \text{I} = x)$ to indicate the probability that actor x attaches to the event that role II is filled by actor y given that actor x is filling role I. Similarly, $p_{xy} = \text{prob}(\text{I} = x | \text{II} = y)$. The actors' beliefs are specified in figure 3.2.

Mathematical huffing and puffing is not necessary to analyze the situation. If actor c gets to play, then she thinks it highly likely that role I is occupied by actor b. Her pure strategy 2 is so adverse in this eventuality[3] that she will always play her pure strategy 1. Thus, if actor a gets to play, he will choose his pure strategy 1 because he then believes it is certain that role II is occupied by actor c. If actor d gets to play, then she believes it is certain that actor b is occupying role I. Hence she will choose her pure strategy 2. Finally, consider actor b. Given our analysis of the plans of the actors c and d, it is optimal for him to play his first pure strategy whatever he believes.

The outcomes which result from this analysis are circled in figure 3.1. Note the dramatic improvement in actor a's payoffs compared with those obtained when the identity of the actors is revealed after their selection. This can be attributed to the fact that, if actor a is chosen to fill role I, then he *knows* that role II is occupied by actor c. He also knows that actor c is badly mistaken in assigning probability 9/10 to the event that role I is occupied by actor b. Thus actor a is in a position to exploit actor c's ignorance and does so.

This completes the analysis of the example. But some further comment will lead us to the next point. Note that actor a does *not* play according to the maximin principle, even though it is common knowledge that the payoffs in the final outcome always sum to zero. Given that his beliefs rule out the possibility that role II is occupied by actor d, his maximin choice is pure strategy 2. But his rational choice, as explained in example 3.1.1, is pure strategy 1. This should not be too surprising because we are studying the situation as a *four*-player game. If the actors a, b, c, and d choose mixed strategies $\boldsymbol{\alpha}$, $\boldsymbol{\beta}$, γ, and $\boldsymbol{\delta}$ respectively in this game, then their expected payoffs are

$$\phi_a(\boldsymbol{\alpha}, \boldsymbol{\beta}, \gamma, \boldsymbol{\delta}) = \pi_a(p_{ca}\boldsymbol{\alpha}^{\mathrm{T}} A_{ac}\gamma + p_{da}\boldsymbol{\alpha}^{\mathrm{T}} A_{ad}\boldsymbol{\delta})$$

$$\phi_b(\boldsymbol{\alpha}, \boldsymbol{\beta}, \gamma, \boldsymbol{\delta}) = \pi_b(p_{cb}\boldsymbol{\beta}^{\mathrm{T}} A_{bc}\gamma + p_{db}\boldsymbol{\beta}^{\mathrm{T}} A_{bd}\boldsymbol{\delta})$$

$$\phi_c(\boldsymbol{\alpha}, \boldsymbol{\beta}, \gamma, \boldsymbol{\delta}) = \pi_c(-p_{ac}\boldsymbol{\alpha}^{\mathrm{T}} A_{ac}\gamma - p_{bc}\boldsymbol{\beta}^{\mathrm{T}} A_{bc}\gamma)$$

$$\phi_d(\boldsymbol{\alpha}, \boldsymbol{\beta}, \gamma, \boldsymbol{\delta}) = \pi_d(-p_{ad}\boldsymbol{\alpha}^{\mathrm{T}} A_{ad}\boldsymbol{\delta} - p_{bd}\boldsymbol{\beta}^{\mathrm{T}} A_{bd}\boldsymbol{\delta})$$

where π_a, π_b, π_c, and π_d are the probabilities which a, b, c, and d respectively attach to the event of their being selected to play before the selection is made. (Recall that unselected actors get a zero payoff.) Nothing says that Nash equilibrium strategies in an n-player game with $n \geqslant 3$ need to be security strategies even if the game is zero-sum. Indeed, the three-player game odd-man-out shows otherwise. Each player simultaneously shows "heads" or "tails." If there is an "odd man out," he pays the other players one unit each: if not, no payments are made. A security strategy requires heads and tails to be shown with equal probability. But (head, head, head) is a Nash equilibrium.

Not only do we have a four-player game to deal with, but there is no guarantee that this game is zero-sum, even though it is common knowledge that the payoffs in the final outcome will sum to zero. This seems paradoxical until one takes into account the fact that nothing has been said about the *consistency* of the actors' beliefs. If John believes that a coin will fall heads with probability 3/4 and Mary believes that it will fall tails with probability 3/4, then they have a good bargain in prospect. John may agree to pay Mary \$1 if tails appears provided that Mary agrees to pay John \$1 if heads appears. The final payoffs resulting from this deal will necessarily sum to zero. But John's *expected* payoff is $1/2 = (+1) \times 3/4 + (-1) \times 1/4$ and Mary's *expected* payoff is also $1/2 = (-1) \times 1/4 + (+1) \times 3/4$. These expectations do not sum to zero.

We have seized the opportunity to make this point because it is often naively proposed that bargaining problems can be solved by appealing to random events about which the players hold different beliefs. For example, if John and Mary were in dispute on how to divide a cake between them, the existence of the coin described in the previous paragraph would effortlessly resolve their difficulty. They could agree that John would receive the whole cake if heads appeared more often than tails in 1,001 tosses of the coin and that Mary would receive the whole cake otherwise. Both would then congratulate themselves on being nearly certain of getting the whole cake.

But can rational players "agree to disagree" in this manner? Aumann (1976) convincingly argues otherwise. If Mary is a rational person, then she will hold her beliefs for rational reasons. If her beliefs differ from John's and John is also rational, then this is presumably because her data differs from John's. But, as soon as Mary's beliefs are revealed to John, then John has access to Mary's data (in so far as this is summarized in her beliefs). John should then revise his own beliefs to take into account this new information. Mary should do the same. Under appropriate conditions, Aumann (1976) then shows that their revised beliefs will necessarily be the same.

For beliefs to be consistent in example 3.1.1, we require that all actors attach the *same* probability q_{xy} to the event that any particular pair (x, y) of actors is selected to play the game. We then have, for example, that

$$p_{ca} = \text{prob}(\text{II} = c \mid \text{I} = a) = \frac{\text{prob}(\text{II} = c \text{ and } \text{I} = a)}{\text{prob}(\text{I} = a)} = \frac{q_{ac}}{\pi_a}$$

and hence $\pi_a p_{ca} = q_{ac}$. Similarly $\pi_c p_{ac} = q_{ac}$. Inserting these results into our formulae for ϕ_a, ϕ_b, ϕ_c, and ϕ_d, we obtain that

$$\phi_a + \phi_b + \phi_c + \phi_d = 0$$

for all $(\boldsymbol{\alpha}, \boldsymbol{\beta}, \gamma, \boldsymbol{\delta})$. Thus the game is zero-sum provided that the beliefs are consistent.

As it happens, the beliefs in example 3.1.1 *are* consistent. Suppose that it is common knowledge that the probabilities q_{xy} with which the pairs (x, y) are chosen is as given in figure 3.3.

Then,

$$p_{ca} = \frac{q_{ac}}{q_{ac} + q_{ad}} = 1 \qquad p_{da} = 0$$

$$p_{cb} = \frac{q_{bc}}{q_{bc} + q_{bd}} = \frac{9}{99} \qquad p_{db} = \frac{90}{99}$$

$$p_{ac} = \frac{q_{ac}}{q_{ac} + q_{bc}} = \frac{1}{10} \qquad p_{bc} = \frac{9}{10}$$

$$p_{ad} = \frac{q_{ad}}{q_{ad} + q_{bd}} = 0 \qquad p_{bd} = 1$$

precisely as given in figure 3.2. Thus we have been dealing with a four-player zero-sum game.

This long discussion would be inexcusable if its only purpose were to make a point about the nature of zero-sum games. But the zero-sum question just provides a convenient hook on which to hang a more fundamental discussion of common knowledge and consistency. These issues need to be thought about quite carefully if confusion is to be avoided.

Example 3.1.2

Another approach to the game of example 3.1.1 is possible. Suppose that the actors chosen to play the game do not know game theory and hence seek advice from game theorists on what to do. Imagine that the actor occupying the role of player I consults Von Neumann and the actor occupying the role of player II consults Morgenstern. In return for their advice, Von Neumann and Morgenstern are to receive 10 percent of the payoff (positive or negative) that the actor they advise finally receives. Their interests are then identical with those of the actor they advise. Having specified Von Neumann and Morgenstern's *tastes*, we also need to specify their *beliefs*. Our assumption will be that the table of figure 3.3 is common knowledge. What we are then left with is a *two-player* game in which Von Neumann is player I and Morgenstern is player II. The actors are now to be treated just as puppets who simply do what they are told.

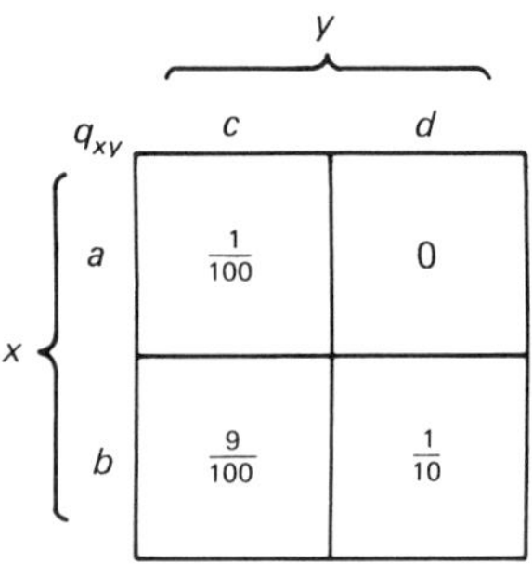

Figure 3.3

What are the strategies for Von Neumann (player I) and Morgenstern (player II)? Von Neumann has *four* pure strategies. These consist of four functions or *decision rules*

$$f_{ij}\colon \{a, b\} \to \{1, 2\} \qquad (i = 1, 2; j = 1, 2)$$

which are to be offered as advice to his client. We list the decision rules f_{11}, f_{12}, f_{21}, and f_{22} as tables in figure 3.4. If the advice f_{ij} is offered, this means

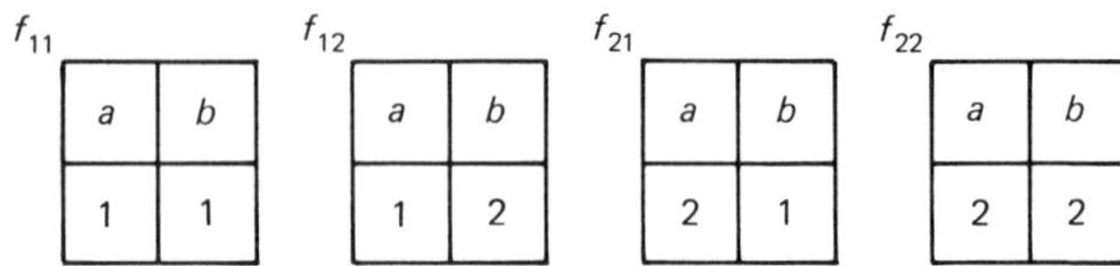

Figure 3.4

		II		
	g_{11}	g_{12}	g_{21}	g_{22}
I f_{11}	+296	−424	+460	−260
I f_{12}	−100	−820	+136	−584
I f_{21}	+288	−432	+464	−256
I f_{22}	−108	−828	+140	−580

Figure 3.5

that the client is to play his pure strategy i if he is an actor of type a and his pure strategy j if he is an actor of type b.

Similarly, Morgenstern has four pure strategies which are identified with the decision rules

$$g_{ij}\colon \{c, d\} \to \{1, 2\} \qquad (i = 1, 2; j = 1, 2)$$

Von Neumann's payoffs are proportional to those given in the 4×4 matrix of figure 3.5. Since the game is zero-sum, these correspond to negative payoffs for Morgenstern.

The calculation for the entry corresponding to (f_{12}, g_{22}), for example, is

$$-5.84 = (-8) \times \frac{1}{100} + 8 \times 0 + 16 \times \frac{9}{100} + (-8) \times \frac{90}{100}$$

It is worth checking some of the other entries if only to get some feeling for how tiresome such calculations are.

Observe that pure strategy g_{12} strongly dominates pure strategies g_{11}, g_{21}, and g_{22}. After the latter have been deleted, pure strategy f_{11} strongly dominates f_{12}, f_{21}, and f_{22}. The solution of this game is therefore the pure strategy (f_{11}, g_{12}). Thus, if Von Neumann and Morgenstern's advice is followed, the actors will behave precisely as in example 3.1.1.

This concludes example 3.1.2 but it is instructive to explore why the same answer is obtained as in example 3.1.1. For this purpose, we require the game tree of figure 3.6 (see section 2.2). The outcome payoffs are taken from the

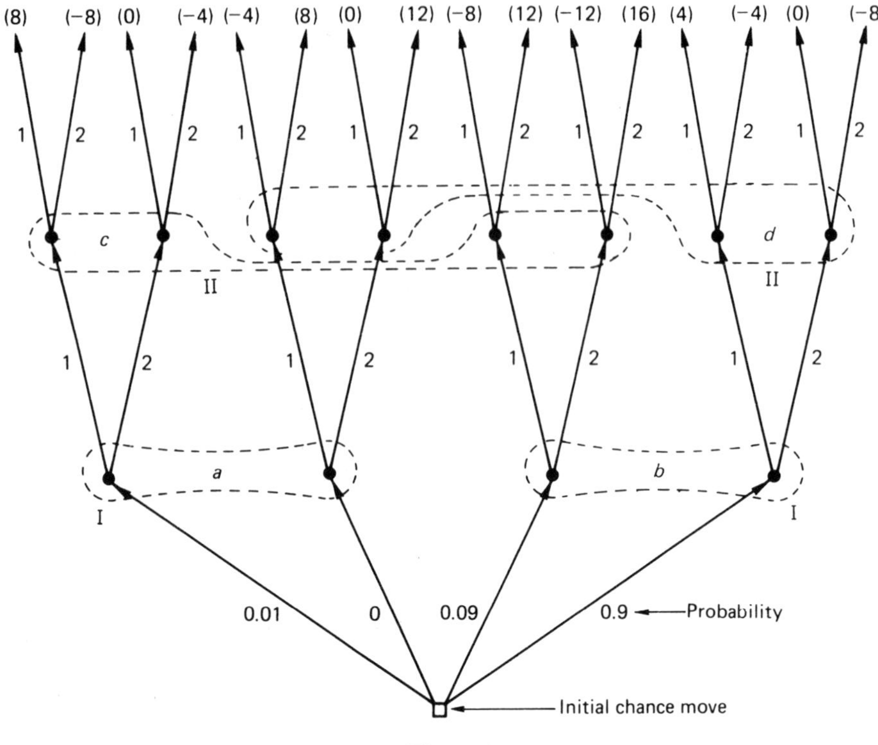

Figure 3.6

payoff tables of figure 3.1. These are positive payoffs to player I and negative payoffs to player II. The probabilities attached to the initial chance move are taken from the table of figure 3.3.

The game tree shows player I learning his type (*a* or *b*) after this has been selected by the initial chance move. (Or, if you prefer, it shows Von Neumann's client telling Von Neumann all about himself.) Player I then selects pure strategy 1 or pure strategy 2 *in ignorance* of the type of player II: hence the information sets marked *a* and *b*. Next player II learns his type (*c* or *d*) and selects pure strategy 1 or 2 *in ignorance* of the type of player I and of the pure strategy selected by player I: hence the information sets marked *c* and *d*.

The point here is that it does not matter whether player I makes his plans about what to do before or after the initial chance move.[4] If he decides *before* the chance move, then his plans will be contingent on what information set (*a* or *b*) he finds himself at after the chance move. A pure strategy choice in this situation will therefore consist of one of the functions f_{11}, f_{12}, f_{21}, or f_{22} of example 3.1.2. If he (and player II) wait until *after* the chance move to decide, then this reduces the situation to the four-player game studied in example 3.1.1, in which the four "players" are the possible actors or types a, b, c, and d.

For this reason, it is possible to retain the terminology of example 3.1.2 while employing the less tiresome calculation technique of example 3.1.1. When proceeding in this way, it is usual to assert that players I and II will choose a *Bayesian equilibrium* (alternatively called a Bayes–Nash equilibrium) in example 3.1.2. This idea is explained in the next section. Those entirely new to the subject are advised to leave this for a second reading.

However, before figure 3.6 fades entirely from memory, there is a final point which is worth making concerning the maximin principle. Von Neumann and Morgenstern's optimal strategies f_{12} and g_{22} are security strategies for the game of figure 3.5 because this is zero-sum. But the behavior we have been advocating in figure 3.6 is *not* explicable as maximin behavior. The reason is that, *after* player I learns he is of type a, the use of the maximin principle would require him to choose 2 rather than 1 from the choices available to him at information set a for precisely the same reason that pure strategy 2 is the security strategy for actor a in the four-player game of example 3.1.1. But player I's optimal choice at information set a is to choose 1. Thus the maximin principle is not even defensible for two-person zero-sum games when these have some dynamic structure (see Aumann and Maschler, 1972).

3.2 Bayesian ideas

To define a *Bayesian equilibrium* in the context of the examples studied in section 3.1, we first need to compute the conditional payoff functions $\phi(f, g|\text{I}=x)$ for all pairs (f, g) of decision rules and for all possible types x of player I. For example,

$$\phi(f_{12}, g_{22}|\text{I}=b) = 16 \times \frac{9}{99} + (-8) \times \frac{90}{99}$$

We also need the corresponding conditional payoff functions for player II. A pair $(\tilde{f}, \tilde{g})$ is then a *Bayesian equilibrium* if and only if

$$\phi(\tilde{f}, \tilde{g}|\text{I}=x) \geqslant \phi(f, \tilde{g}|\text{I}=x)$$

$$\phi(\tilde{f}, \tilde{g}|\text{II}=y) \geqslant \phi(\tilde{f}, g|\text{II}=y)$$

for *all* pairs (f, g) of decision rules and for *all* pairs (x, y) of types. What these inequalities are telling us is that player I and player II will never want to change their plans after they learn their types. This is no big deal. All that is being said is that it will never profit any actor to deviate from the advice given by Von Neumann or Morgenstern unless some other actor deviates also. But Von Neumann and Morgenstern's advice would not be optimal otherwise. To say that (f_{12}, g_{22}) is a Bayesian equilibrium says no more than that (f_{12}, g_{22}) is a Nash equilibrium in the game between Von Neumann and Morgenstern described in example 3.1.2 or, equivalently, that the pure strategy profile (1, 2, 2, 2) is a Nash equilibrium for the four-player game described in example 3.1.1.

Why introduce the idea of a Bayesian equilibrium if it only expresses ideas already built into a Nash equilibrium? The reader is referred to Harsanyi (1967–8) or Myerson (1979) for the appropriate philosophy. However, we avoid the terminology for several reasons. Firstly, it is always possible to say whatever is necessary without anything more than the idea of a Nash equilibrium. Secondly, there are difficulties with the interpretation of Bayesian equilibrium if this is taken as a primitive notion.[5] Thirdly, the word "Bayesian" itself represents an unwelcome distraction.

This last observation provides an opportunity for an aside on "Bayesianism" in general. It might be thought that to describe something as "Bayesian" was to imply some close connection with the works of Thomas Bayes. But, as the example of a Bayesian equilibrium demonstrates, this is not the case. A Bayesian is *nominally* just a theorist who models decision making in terms of the maximization of expected utility relative to a subjective probability distribution. In the language of the preceding section, a decision-maker's utility function describes his *tastes* and his subjective probability distribution describes his *beliefs*. Subjective probability ideas were developed by De Finetti (1974) and Ramsay (1931), among others, and unified with the ideas of Von Neumann and Morgenstern (1944) on utility function in Savage's (1954) *Foundations of Statistics*. Savage's theory is entirely and exclusively a *consistency* theory. It has nothing to say about *how* decision-makers come to have the tastes and beliefs ascribed to them; it asserts only that, if the decisions taken by the decision-makers are consistent (in a sense made precise by a list of formal "axioms"), then they act *as though* they maximize expected utility relative to subjective probabilities. Objections to the axiom system can be made, although it is no objection to argue that real people are often inconsistent in the sense specified by the axioms. People also often get their sums wrong, but we would not therefore argue for a change in the axiomatic foundations of arithmetic. However, whatever doubts there may be about the foundations of Savage's theory, this chapter treats the theory as gospel. To this extent, the chapter is based on the "Bayesian" principles.

Section 5 of chapter 5 seeks to explain why such Bayesian principles do not in themselves provide an adequate foundation for game theory. The essential reason is that "Bayesian rationality" is concerned only with the *consistency* of the beliefs held by players. It has nothing to say about the *origins* of such systems of beliefs. Naive Bayesians proceed as though Bayesian rationality endows its fortunate adherents with the capacity to "pluck their beliefs from the air," but this is not an approach that can lead to anything that is very useful in a game-theoretic context. Rational players need to *construct* the beliefs they hold in equilibrium by means of "If I think that you think that I think ..." chains of reasoning. An attitude of mind that directs attention away from such underlying *tâtonnement* processes would seem to make nonsense of the meaning implicit in the word "equilibrium."

3.3 Trembling motives

In section 2.9, we discussed the stability of equilibria in the presence of small perturbations in the parameters of the game. Our concern there was with players whose aims were unambiguous but who had difficulty in carrying out those aims as a result of the intervention of small random errors between the decision to act and the act itself. The players were said to have "trembling hands." As they reached to play P–K4, there was always the possibility that they might play P–KR4 by mistake.

Rather than studying such "trembling hands," Harsanyi (1973) studied "trembling motives." This is a useful topic to introduce at this point, not only because it ties together the theory of incomplete information and the idea of stability, but also because it sheds light on mixed strategies. Recall that interpretations of mixed strategies were given in section 2.6 that do not require us to envisage individual agents actually randomizing. The context was that of a *tâtonnement* involving myopic players. The following model of Harsanyi allows a similar "purification" of the notion of a mixed strategy even when the players are conscious optimizers.

The twist is that the players are assumed not to be as fully informed about the motives of their opponents as they would like to be. This does *not* mean that we are about to study situations in which players are uncertain about that their opponents would do when offered a choice between $100 and nothing. But there may very well be uncertainty about the precise attitude of an opponent to the bearing of risk. Indeed, a particular individual's attitude may vary slightly from day to day. Such variations will be reflected in the values of the player's Von Neumann and Morgenstern utilities and hence in the payoffs of the game.

Another source of uncertainty might relate to the beliefs of the opponent. In poker, for example, it is possible that two players might have slightly different beliefs about the probabilities with which given hands are dealt.[6] This will lead them to evaluate the probabilities with which a given hand will win the pot differently, and hence cause them to attach different payoffs to the same strategy profiles.

An incomplete information model is necessary to deal with such random fluctuations in the payoffs. As the basis for an appropriate example, we take the zero-sum game with matrix

$$A = \begin{pmatrix} 1 & 4 \\ 3 & 2 \end{pmatrix}$$

studied already in example 2.3.5.

First, two independent chance moves select a pair (E, F) of 2×2 error matrices. The matrices E and F represent random fluctuations in the preferences of players I and II. Each actor who may occupy the role of player I is identified with one of the matrices E and each actor who may occupy the role of player II is identified with one of the matrices F. After a pair of actors has been chosen to play (i.e. a pair (E, F) has been selected by the opening chance move), they face a bimatrix game in which the payoff matrix for actor E is $A + E$ and the payoff matrix for actor F is $-(A + F)$. An actor knows his or her own payoff matrix but not that of the other actor. An actor's beliefs about the opponent's payoffs are derived from the probability distributions governing the initial chance moves (about which we shall be more specific shortly).

Each actor E chooses a mixed strategy $\mathbf{p}(E) = (1 - p(E), p(E))$ and each actor F chooses a mixed strategy $\mathbf{q}(F) = (1 - q(F), q(F))$. Actor E then receives an expected payoff of

$$\phi_{\mathrm{E}}(\mathbf{p}, \mathbf{q}) = \int \mathbf{p}(E)^{\mathrm{T}}(A + E)\mathbf{q}(F)\, \mathrm{d}\mu(F) = \mathbf{p}(E)^{\mathrm{T}}(A + E)\mathbf{q}_0$$

where the integral extends over all possible F, μ is the probability measure which governs the choice of F, and

$$\mathbf{q}_0 = \int \mathbf{q}(F)\, \mathrm{d}\mu(F)$$

is the expected value of $\mathbf{q}(F)$. In choosing an optimal response, actor E therefore behaves as though he were facing a single opponent who plays $\mathbf{q}_0$ (rather than a population of actors who play in various different ways). We can therefore borrow the optimal response diagram 2.7(b) of example 2.3.5, provided that this is adapted to take account of the fact that A is replaced by $A + E$. In the adapted diagram, it will not be true that actor E is indifferent between all his mixed strategies when $q_0 = 1/2$. Instead, this indifference will occur at some other value $q_0 = q_1(E)$. Recall, however, that we are interested in the case when uncertainties are small. We interpret this to mean that there exists a small number $\delta > 0$ with the property that the absolute value of each entry of E is less than δ. But, if δ is small, the adapted optimal response diagram for $A + E$ will be approximately the same as that for A (in the sense that, for each $\varepsilon > 0$, there exists a $\delta_0 > 0$ such that $|q_1(E) - 1/2| < \varepsilon$ provided that $\delta < \delta_0$). In particular, $q_1(E)$ will be approximately $1/2$. Similar

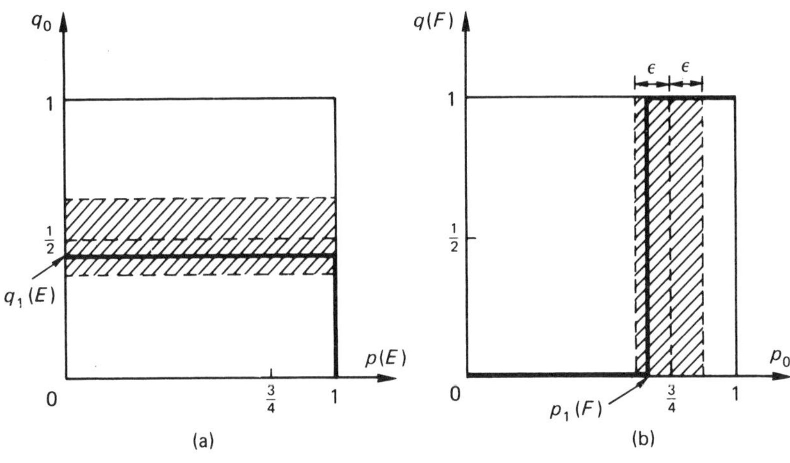

Figure 3.7

considerations apply to actor F. Figure 3.7(a) shows an adapted optimal response diagram for actor E and figure 3.7(b) shows a corresponding diagram for actor F.

Suppose that Nash equilibrium strategies for actors E and F are $\tilde{p}(E)$ and $\tilde{q}(F)$. Note several things. Firstly, unless it happens that $\tilde{q}_0 = q_1(E)$, then actor E will choose a *pure* strategy (i.e. $\tilde{p}(E) = 0$ or $\tilde{p}(E) = 1$). If the probability measure λ for E has a probability density function, it follows that *almost all* actors E choose a pure strategy (i.e. player I chooses a pure strategy with probability 1). Similar considerations, of course, apply to player II. Next observe that, if $\tilde{q}_0$ is not approximately 1/2 (i.e $|\tilde{q}_0 - 1/2| \geqslant \varepsilon$), then all actors E will choose the *same* pure strategy. It then follows that $\tilde{p}_0$ will not be approximately 3/4 and so all actors F will choose the *same* pure strategy. But there cannot be a Nash equilibrium in which all actors in the same role choose the same pure strategy for the same reason that the original game with matrix A has no pure strategy equilibria. It follows that $\tilde{p}_0$ is approximately 1/2 and $\tilde{q}_0$ is approximately 3/4.

The point here is that, although players I and II both choose pure strategies almost always, the small random fluctuations in the player's preferences mean that the pure strategy the other player will use cannot be predicted precisely. Instead, each player will have to assign a probability to the event that the opponent will play pure strategy 2. These probabilities $\tilde{p}_0$ and $\tilde{q}_0$ are approximately the probabilities which arise when the mixed strategy Nash equilibrium is calculated for the original game (with matrix A and no random fluctuations). To broaden the point further, a mixed strategy does not have to be seen as an active random choice by a player i: it can also be seen as measuring the uncertainty inside the head of the opponent about what *pure* strategy player i will choose.

3.4 Revelation

Much of the game theory literature that deals with incomplete information is concerned with the design of "incentive-compatible mechanisms." Consider, for example, a benevolent government whose aim is to implement an ideal tax and benefits system. Such a system may require distinguishing between the idle-rich and the poor-but-honest. But then the question of deceit arises. If, for example, the only way the government could collect data on its citizens were by distributing questionnaires, it would not help simply to ask, "Are you poor-but-honest?" If the answer "yes" results in a benefit check and the answer "no" in a tax demand, the idle-rich have no incentive to reply truthfully. A system of taxing the idle-rich to subsidize the poor-but-honest would therefore not be *implementable* because the government would have no way of distinguishing between these two groups.

To be implementable, a system which depends on data gathered from the individuals to which it is to be applied needs to incorporate incentives which encourage the individuals to reveal the necessary data. It does not matter *how* this data is revealed. It may be, for example, that the data is revealed only *indirectly* as a consequence of inferences being drawn from the individual's behavior even though this behavior is intended to be misleading.

If a system is implementable in this sense, the arrangement which implements it is called an *incentive-compatible mechanism.* Such a mechanism will form the script for a game. A *static* mechanism assigns a message space M_i to each player $i = 1, 2, \ldots, n$. A function $f: M_1 \times M_2 \times \ldots \times M_n \to \Omega$, which everybody knows, then fixes the final outcome $f(m_1, m_2, \ldots, m_n)$ in terms of the messages $m_1, m_2, \ldots, m_n$ which the players send. What messages the players send will depend on their types. As always, the type of a player is determined by his or her tastes (concerning the outcomes in the space Ω) and his or her beliefs (concerning the distribution of types).

A Vickrey (or second-price) auction provides a simple example. An object is to be sold at an auction sale to one of a number of bidders whose valuation of the object is not known to anyone but themselves. There is no point in the auctioneer asking each bidder for his true valuation with the aim of selling to the bidder whose valuation is greatest. The bidders will not necessarily tell the truth. Instead, our auctioneer asks each bidder to submit a sealed bid and undertakes to sell the object to the highest bidder but *at the price bid by the second-highest bidder.*[7] This is not necessarily the optimal auction that an auctioneer anxious to maximize the sale price might design. This will depend on the probabilities which the auctioneer assigns to the various possible types. The interest in a Vickrey auction is that *truth-telling* (i.e. reporting a correct valuation) is a Nash equilibrium for the actors. The highest bid b made by someone else is what I shall have to pay for the object if I win the auction. If my valuation v exceeds b, I want to win the auction and a bid of v ensures that I do. If my valuation v is less than b, then I do not want to win the auction and a bid of v ensures that I do not. If $v=b$, I am indifferent between winning and not winning and so anything (including a bid of v) is optimal.

The *revelation principle* (originated by Gibbard, 1973, and developed by Myerson, 1979, and others) provides a pleasing application of the theory of incomplete information in this context. A static mechanism is a script (or a game form) for a game. We look at the case $n=2$ so that we have only two roles as in section 3.1. We suppose that the actors who may be selected to fill role I are chosen from a population A and those who may fill the role of player II are chosen from a population B. Each actor $a \in A$ chooses a strategy $m_1(a) \in M_1$ and each actor $b \in B$ chooses a strategy $m_2(b) \in M_2$. These are the messages the actors will send if they find themselves called upon to play. For $\tilde{m}_1\colon A \to M_1$ and $\tilde{m}_2\colon B \to M_2$ to describe a Nash equilibrium, we require that no actor should wish to switch his choice of a message given the choices of the other actors. Thus,

$$\int_B \phi_a \circ f(\tilde{m}_1(a), \tilde{m}_2(b))\, \mathrm{d}\mu_a(b) \geqslant \int_B \phi_a \circ f(m_1, \tilde{m}_2(b))\, \mathrm{d}\mu_a(b)$$

for all $a \in A$ and all $m_1 \in M_1$. Similarly,

$$\int_A \phi_b \circ f(\tilde{m}_1(a), \tilde{m}_2(b))\, \mathrm{d}\mu_b(a) \geqslant \int_A \phi_b \circ f(\tilde{m}_1(a), m_2)\, \mathrm{d}\mu_b(a)$$

for all $b \in B$ and all $m_2 \in M_2$. In these inequalities, $\phi_a\colon \Omega \to \mathbb{R}$ and $\phi_b\colon \Omega \to \mathbb{R}$ denote the actors' utility functions and μ_a and μ_b denote the probability distributions which describe their beliefs. In particular,

$$\phi_a \circ f(m_1, m_2) = \phi_a(f(m_1, m_2))$$

is the utility that actor a will derive if he is selected for role I and chooses message m_1 and whoever occupies role II chooses message m_2.

Now look at a second mechanism in which the message spaces are just A and B and the function $F\colon A \times B \to \Omega$ which determines the final outcome is given by

$$F(a, b) = f(\tilde{m}_1(a), \tilde{m}_2(b))$$

This mechanism calls upon the actors to say who they are (i.e. for an actor to reveal his or her type). The revelation principle is the simple observation that truth-telling is a Nash equilibrium in this second mechanism (called a *direct* mechanism because players are asked directly for the required data). The proof is immediate. Replace m_1 by $m_1(\alpha)$ in the first of the preceding inequalities. Then

$$\int_B \phi_a \circ F(a, b)\, d\mu_a(b) \geqslant \int_B \phi_a \circ F(\alpha, b)\, d\mu_a(b)$$

for all $a \in A$ and all $\alpha \in A$. Thus no actor a has an incentive to switch from the announcement that he is actor a provided that all other actors tell the truth as well. Similarly, no actor b has an incentive to lie unless other actors lie also.

What has been shown is that any distribution of payoffs which can be achieved as the result of equilibrium play using an indirect mechanism can also be achieved as the result of a *truth-telling* Nash equilibrium in a *direct* mechanism. Does this mean that planners can simply send out a circular which says, "Tell me what I want to know: things have been fixed so that, if everybody replies truthfully, nobody will have cause to regret it"? This would be a naive conclusion, even if there were no difficulties about the planner's credibility rating.

The problem lies in the fact that the truth-telling equilibrium may be only one of many Nash equilibria. For two-person zero-sum games, this would not be a difficulty since all Nash equilibrium in such games are equivalent and interchangeable (section 2.8). But, in the general case, it matters which equilibrium is selected and the truth-telling equilibrium need not be the most attractive of those available. The Vickrey auction game provides an example. Apart from the truth-telling equilibrium, there are other equilibria. Suppose that everybody in a population of artlovers has a different valuation of a certain art object (which cannot be resold after it has been bought). Precisely two members of the population find themselves in the auction room by chance at the time the sale takes place. They submit sealed bids and the object goes to the high bidder at the price submitted by the low bidder. As explained earlier, it is a Nash equilibrium for each actor to plan to submit his or her true valuation.

However, the truth-telling equilibrium has a rival. Suppose that each artlover plans to submit, not his or her true valuation, but that of the artlover with the next lowest valuation. (The artlover with the lowest valuation of all submits a zero bid.) The result is a Nash equilibrium. The higher bidder certainly has no incentive to change her bid. But nor does the low bidder. If he raises his bid and thereby gets the object, he will have to pay the opponent's bid. This will yield him a payoff of at most zero since the high bid is at least as large as the low bidder's true valuation. Hence the low bidder might as well stick with his current bid.

Note that the lying equilibrium *weakly Pareto dominates* the truth-telling equilibrium. This means that the former assigns each artlover at least as large an expected payoff as the latter. Or, to be more precise, if u is the vector of expected payoffs for the artlovers in the truth-telling equilibrium and v is the vector of expected payoffs in the lying equilibrium, then $\mathbf{v} < \mathbf{u}$ (example 2.3.3). The lying equilibrium would *strongly Pareto dominate* the truth-telling equilibrium if $\mathbf{v} \ll \mathbf{u}$ (example 2.3.1). This does not hold because the artlover with the lowest valuation of all gets nothing in both cases. (In an infinite population, this individual could be eliminated by allowing artlovers with arbitrarily small valuations.)

Although the lying equilibrium is a Pareto improvement on the truth-telling equilibrium, it is not claimed that the latter should *necessarily* be rejected in favor of the former. Proponents of the truth-telling equilibrium can point out that telling the truth is a weakly dominating strategy for each individual separately (example 2.3.3). This means that telling the truth is at least as good for each individual separately as any other strategy *whatever* the other individuals choose to do. However, the point here is simply that there *is* an issue. The fact that truth-telling is *an* equilibrium is not adequate for the conclusion that truth-telling is the equilibrium that should be selected. Some extra argument to this effect must be brought onto the scene.

One reason for this aside on mechanism design is to make the point that it is not the only interesting activity for game theorists, although it is an activity to which a substantial proportion of the literature is devoted. My own inclination is to study games-as-they-are rather than games-as-they-ought-to-be. Partly this is because I am uncomfortable with the use of the revelation principle when I can get no feeling for the strategic nature of the direct revelation mechanism, but mostly it is because I am more optimistic than most game theorists about real-world applications of game theory. To deal with games-as-they-are it is necessary to make judgments about the the world *is*. Such judgments can always be criticized, especially since one is often forced into studying games with rules that one knows oversimplify the phenomena under study because, if more realistic rules were employed, the result would be something too complex to analyze. But there are certain criticisms that should be resisted.

One criticism sometimes made of attempts to model games-as-they-are is that their rules are not socially optimal (in some sense) and therefore they are not worthy of study. One should instead study games whose rules have been designed optimally. If the criticism is interpreted naively, it has a simple answer. Perhaps bridge is a better game than whist, but studying bridge will not help much in predicting the behavior of those who play whist.

However, the criticism is sometimes proposed in a more subtle form. It is argued that social institutions evolve just like everything else. It is, after all, true that whist has largely been displaced by bridge. The only interesting institutional arrangements, so the argument goes, are therefore the end-products of such an evolutionary process, and these should be anticipated to be socially optimal. Without conceding the last point, it has to be admitted that this is a difficult criticism for a game theorist to respond to. He usually models humans as rational optimizers and, if pressed, will trot out an evolutionary story in defense of this practice. Why should be balk at a similar story concerning human institutions? I think the proper reply begins with an admission that modeling humans as rational optimizers is unsatisfactory. It is, at best, only an approximation valid only under restricted circumstances. But, having made one unsatisfactory assumption because one does not know what else to do, it does not follows that other unsatisfactory assumptions should be made as well. Human institutions (i.e. the rules of real-life games) are accessible to empirical investigation in a way that the inside of a human head is not. We *know*, not only that all is *not* for the best in this best of all possible worlds: we also have data on how the world actually is. This data ought to be used.

None of this is intended to deny the prescriptive value of optimally designed incentive mechanisms or arbitration schemes. All that is being suggested is that other parts of game theory are at least as interesting. Nor is it suggested that tools from mechanism design cannot be useful elsewhere. For example, the revelation principle can be very useful in demonstrating that certain outcomes *cannot* be obtained as equilibria *however* the game-as-it-is may have been chosen from a given class of games.

3.5 Von Neumann's Poker models

Poker is a game in which information is of paramount importance. It is therefore not surprising that game-theoretic analyses of poker models should be instructive on informational questions. We begin with a brief outline of what goes on at the poker table for those fortunate enough never to have been led astray.

In straight poker, each player begins by placing a small amount (the *ante*) into the *pot*. The cards are then dealt, after which the betting commences. Bets are made clockwise round the table starting from the dealer's left. Each player is required to bet at least as much as the player to his or her right, or else to *fold*. A player who folds loses all he or she has staked and takes no further part in the game untii the next deal. To bet more than the player on your right is to *raise*. To bet the same is to *call*. The betting ends when all the players who have not folded have staked the same amount. These players then show their cards and the best *hand* takes the whole pot.

This basic version of poker is seldom actually played. Innumerable variations exist such as draw, stud (five-card or seven-card), push-em, hi-lo, and the like. Usually several rounds of betting take place between which the players have an opportunity of improving their hands. Limits on the amounts which can be bet are usual but the game becomes dull if the bets cannot rise exponentially up to an amount which it really hurts a player to lose. Pot-limit, for example, confines any raise to at most what is in the pot already.

The "natural" strategy at poker is to bet high with a good hand and to bet low with a bad hand. But this Colonel Blimp strategy is not likely to work well in the long run because the other players will learn to gauge the quality of your hand by the size of your bet. It is therefore necessary to *bluff* sometimes. In its crudest form, this means making a large raise on a poor hand. The aim is not so much that you hope to win with the poor hand by frightening the other players into folding, although such a windfall will gratefully be received. The aim is to advertise that a big bet from you does not necessarily mean that you have a good hand. Thus, when you *are* dealt a good hand and bet high, the opposition will not automatically fold and you will therefore have some prospect of making a killing.

The question is: how often should one bluff in order that the extra profits garnered with good hands should outweigh the losses to be expected from betting high on some bad hands? The answer is: a great deal more than most people think at all likely.

Von Neumann's poker models are highly simplified even compared with straight poker. For one thing, the game is played once and once only so that the question of learning over time is not an issue. Moreover, only two players are involved. Nevertheless, as parables on optimal bluffing levels, they are very instructive – especially in respect of player I's behavior in model 2.

3.5.1 Model 1

Each of two players antes $\$a$ and is then dealt a card from a deck consisting of all real numbers in the compact interval $[0, 1]$. The values of the cards are independent of each other and uniformly distributed on $[0, 1]$. We denote the card dealt to player I by u and that dealt to player II by v. Having seen their own cards, but *not* their opponent's, each player then *simultaneously* bets either low ($0) or high ($\$b$, where $b > 0$). This concludes the betting. Both players now either have the same amount in the pot or else one has less than the other. In the former case, the players are deemed to have "called" and they show their hands, the player with the larger card[8] scooping the pot. In the latter case, the player with less money in the pot is deemed to have "folded" and the other player takes the pot *regardless* of who has the larger card. Figure 3.8 shows the appropriate payoff tables. As always in zero-sum games, it is the payoff to player I which is recorded.

We propose to analyze the game like a game of incomplete information. The type of an actor will be determined by the card he or she is holding. The larger the card, the more probable it is that the actor will win if there is a showdown. The actor's type therefore affects his or her preferences over

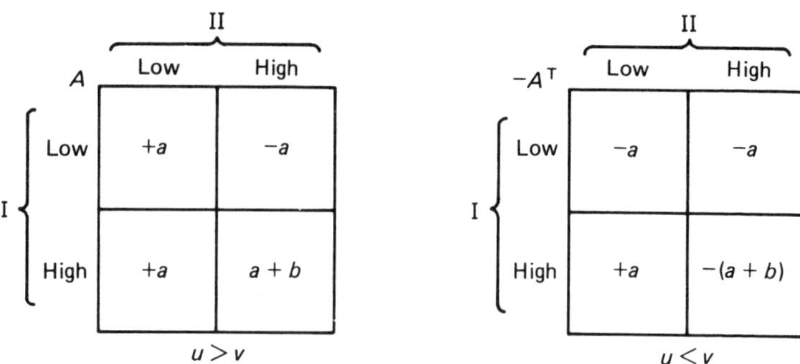

Figure 3.8

strategy profiles. However, all actors are assumed to have preferences which are "linear in money," i.e. the Von Neumann and Morgenstern utility of $\$x$ can simply be taken to be x.

Suppose that actor u (player I holding card u) uses mixed strategy $\mathbf{p} = (1 - p(u), p(u))$ and that actor v (player II holding card v) uses mixed strategy $\mathbf{q} = (1 - q(v), q(v))$. A necessary and sufficient condition[9] for a Nash equilibrium in the game played by the actors is that $p = q = \tilde{r}$, where the function $\tilde{r}$: $[0, 1] \to [0, 1]$ is as illustrated in figure 3.9(a). Note the bluffing zone where players sometimes bet high when dealt a bad card. Note also the sharp cut-off point between the bluffing zone and the rest of the range. The reasons for bluffing in this zone are also instructive. If player I is dealt a card u in the bluffing zone, then he will be *indifferent* between playing "high"

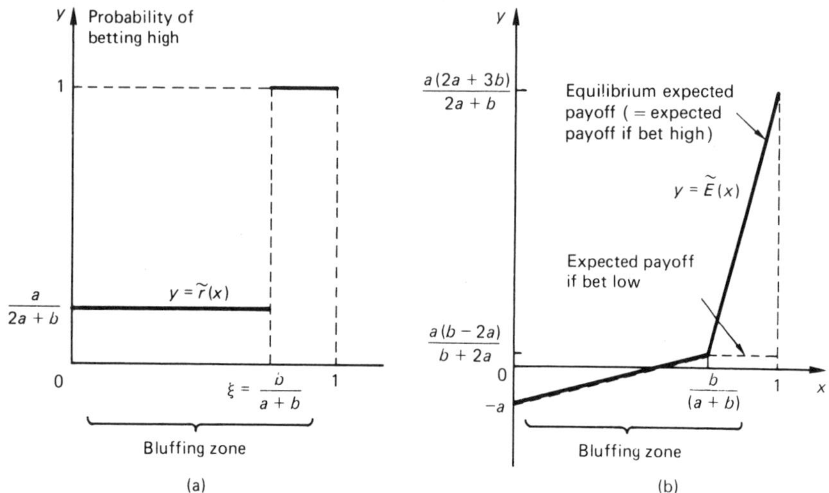

Figure 3.9

and playing "low," given optimal play by the opposition. Otherwise a mixed strategy could not be an optimal response (see ch. 2, n. 11). The bluffing is therefore defensive in that player I does not expect to gain by bluffing. He bluffs because he would be exploited if it were known that he did not bluff.

We shall only check that the function $\tilde{r}$ does determine a Nash equilibrium. A methodology for locating such equilibria is given with model 2. First we need the expected payoff $E(u)$ to actor u, given that $\mathbf{p}$ and $\mathbf{q}$ are used. From figure 3.8, we have that $\mathbf{p}^{\mathrm{T}}A\mathbf{q}$ is the appropriate payoff when it is known that $u > v$, and $\mathbf{p}^{\mathrm{T}}(-A^{\mathrm{T}})\mathbf{q}$ is the appropriate payoff when it is known that $u < v$. Hence

$$
\begin{aligned}
E(u) &= \int_0^u \mathbf{p}^{\mathrm{T}} A\mathbf{q}\, \mathrm{d}v + \int_u^1 \mathbf{p}^{\mathrm{T}}(-A^{\mathrm{T}})\mathbf{q}\, \mathrm{d}v \\
&= p(u)S(u) + T(u)
\end{aligned}
$$

where

$$
S(u) = (2a+b)\left[\int_0^u q(v)\, \mathrm{d}v - \int_u^1 q(v)\, \mathrm{d}v + \frac{2a}{2a+b}(1-u)\right]
$$

$$
T(u) = a\left[-2\int_0^u q(v)\, \mathrm{d}v + 2u - 1\right]
$$

All that matters for choosing an optimal value of $p(u)$ is whether $S(u) > 0$, $S(u) = 0$, or $S(u) < 0$. If $S(u) > 0$, $p(u) = 1$ is the optimal choice for actor u. If $S(u) = 0$, any choice of $p(u)$ is optimal (including $p(u) = a/(2a+b)$). If $S(u) < 0$, $p(u) = 0$ is optimal. (The value of $T(u)$ is immaterial to what value of $p(u)$ is optimal but is still needed to work out what the payoffs which result from optimal play are.)

When $q = \tilde{r}$ and $\xi = b/(a+b)$, we have that $S(0) = S(\xi) = 0$ and $S(1) = 2a > 0$. Since S is affine (i.e. has a straight-line graph) when restricted to the intervals $[0,\xi]$ and $[\xi,1]$, we conclude that $S(x) = 0$ for $0 \leqslant x \leqslant \xi$ and $S(x) > 0$ for $\xi < x \leqslant 1$. It follows immediately that an optimal response to $q = \tilde{r}$ is $p(u) = \tilde{r}(u)$. By symmetry, an optimal response to $p = \tilde{r}$ is $q(v) = \tilde{r}(v)$. By playing according to $\tilde{r}$, all actors therefore respond optimally to the choices of the others and hence we have a Nash equilibrium.

Note, before leaving the model, that it is a two-person zero-sum game for players I and II *before* they are dealt their cards. It follows that their security levels v_1 and v_2 satisfy $v_1 + v_2 = 0$. As the game is symmetric, $v_1 = v_2$ and so $v_1 = v_2 = 0$. This may be checked by observing that the equilibrium payoff function $\tilde{E}$ of figure 3.8(b) satisfies

$$v_1 = \int_0^1 \tilde{E}(x)\, dx = 0$$

3.5.2 *Model 2*

Model 2 differs from model 1 in its betting rules. The new rules are meant to capture something of the sequential nature of real-life betting in poker. The deal and the ante (\$$a$) remain the same as before. Player I begins by betting either low (\$0) or high (\$$b$, where $b > 0$). If player I bets high, player II has a choice of low (\$0) or high (\$$b$). If player I bets low, player II *must* bet low (\$0). If the betting concludes with both players having the same amount in the pot, then player II is deemed to have called, and the larger card takes the pot. Otherwise player II will have less money in the pot and so be deemed to have folded. Then player I takes the pot regardless of who holds what card. Figure 3.10 shows the payoffs for player I (which are simultaneously negative payoffs for player II).

As in model 1, we require the expected payoff $E_1(u)$ to actor u (player I holding card u) given that **p** and **q** are used. Since the game is not symmetric, a separate expression is necessary for the expected payoff $E_2(v)$ to actor v (player II holding card v). We have that

$$\begin{aligned} E_1(u) &= \int_0^u \{(1-p)a + p[a(1-q) + (a+b)q]\}\, dv \\ &\quad + \int_u^1 \{(1-p)(-a) + p[a(1-q) - (a+b)q]\}\, dv \\ &= p(u)S_1(u) + T_1(u) \end{aligned}$$

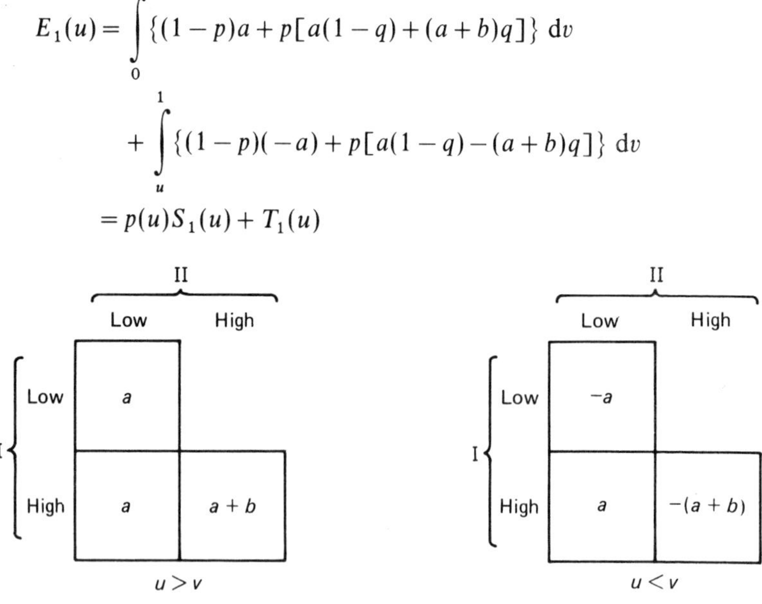

Figure 3.10

where

$$S_1(u) = b\int_0^u q(v)\,\mathrm{d}v - (b+2a)\int_u^1 q(v)\,\mathrm{d}v + 2a(1-u)$$

The value of $T_1(u)$ is irrelevant to the optimal choice of $p(u)$ and so is not reproduced. Similarly,

$$E_2(v) = q(v)S_2(v) + T_2(v)$$

where

$$S_2(v) = -b\int_v^1 p(u)\,\mathrm{d}u + (b+2a)\int_0^v p(u)\,\mathrm{d}u$$

For model 1, we simply plucked some strategies from the air and checked that they constituted a Nash equilibrium. Here we illustrate how Nash equilibrium strategies can be constructed from ground zero.

Suppose that $\tilde{p}(u)$ maximizes $E_1(u)$ when $q = \tilde{q}$ and that $\tilde{q}(v)$ maximizes $E_2(v)$ when $p = \tilde{p}$. Write $\tilde{S}_1$ and $\tilde{S}_2$ for the values obtained by substituting $\tilde{p}$ for p and $\tilde{q}$ for q in the formulae for S_1 and S_2. We then have that $\tilde{S}_1(u) > 0$ implies $\tilde{p}(u) = 1$; $\tilde{S}_1(u) < 0$ implies $\tilde{p}(u) = 0$; and $0 < \tilde{p}(u) < 1$ implies $\tilde{S}_1(u) = 0$. Similar considerations also apply to $\tilde{S}_2(v)$ and $\tilde{q}(v)$.

Observe that $\tilde{S}_2$: $[0,1] \to \mathbb{R}$ is continuous and increasing on $[0,1]$, with $\tilde{S}_2(0) = -bI$ and $\tilde{S}_2(1) = (b+2a)I$, where $I = \int_0^1 \tilde{p}(u)\,du$. These properties are summarized in figure 3.11(a), where $[\xi,\eta]$ is the largest interval on which $\tilde{S}_2(u) = 0$. The properties indicated for $\tilde{q}$ in figure 3.11(b) follow immediately.

Next observe that $\tilde{S}_2(v) = 0$ on any interval if and only if $\tilde{p}(v) = 0$ almost everywhere in that interval. The reason is that $0 = \tilde{S}'_2(v) = -2(b+a)\tilde{p}(v)$ almost everywhere in any interval on which $\tilde{S}_2$ is constant. Thus $[\xi,\eta]$ is the largest interval on which $\tilde{p}(u) = 0$ almost everywhere. It follows that $\tilde{S}_1(\xi) = \tilde{S}_1(\eta) = 0$. Our knowledge of $\tilde{q}$ allows us to calculate $\tilde{S}'_1(u) = -2a$ $(0 < u < \xi)$ and $\tilde{S}'_1(u) = 2b$ $(\eta < u < 1)$. These properties of $\tilde{S}_1$ are incorporated in figure 3.11(c). The characteristics of $\tilde{p}$ illustrated in figure 3.11(d) then follow immediately.

We shall proceed no further with the calculation, since it is the shape of the graph of $\tilde{p}$ on which we wish to focus. Note, however, that the values

$$\xi = \frac{ab}{(a+b)(b+4a)} \qquad \eta = \frac{b^2+4ab+2a^2}{(a+b)(b+4a)}$$

can be obtained from the observation that $\tilde{S}_1(\xi) = \tilde{S}_1(\eta) = \tilde{S}_2(\xi) = \tilde{S}_2(\eta) = 0$. Many equilibrium values of $\tilde{q}$ exist. It turns out that $\tilde{q}$ can be chosen freely

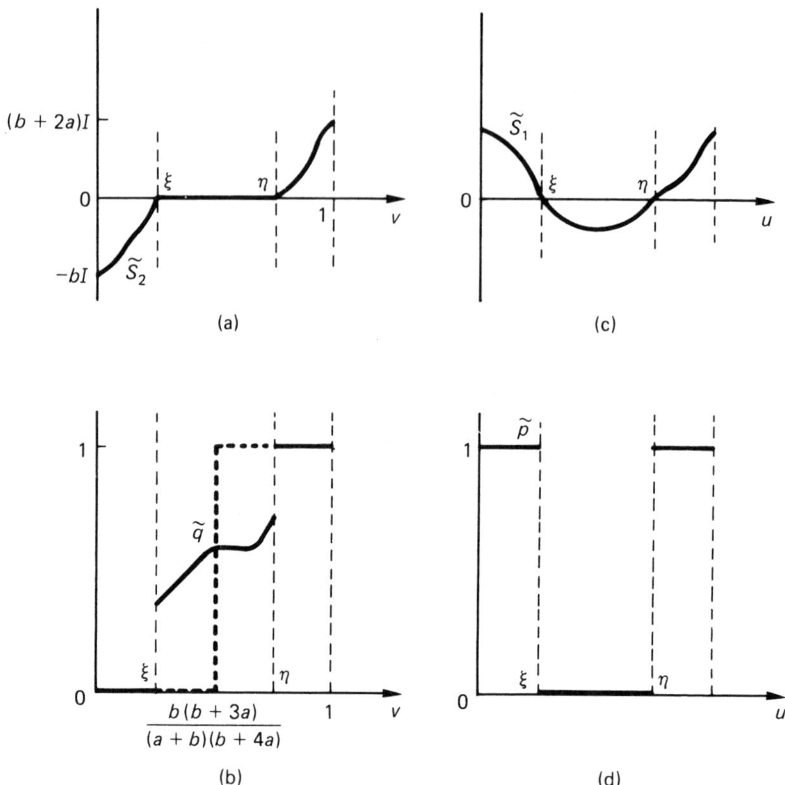

Figure 3.11

on the range $[\xi,\eta]$ subject to the constraints

$$\frac{1}{\eta-\xi}\int_{\xi}^{\eta} q(v)\,\mathrm{d}v = \frac{a}{a+b} \qquad \frac{1}{\eta-u}\int_{u}^{\eta} q(v)\,\mathrm{d}v \geqslant \frac{a}{a+b} \qquad (\xi \leqslant u \leqslant \eta)$$

The simplest choice of q is indicated with a dotted line in figure 3.11(b).

It is remarkable that this poker model has *pure strategy* equilibria. At such equilibria, everybody plays deterministically once they have received their cards. To the extent that their behavior seems random to an observer, this will be because of the randomness of the deal. It is perhaps not so surprising that player II should have a Colonel Blimp equilibrium pure strategy in which she bets high with a high card and low with a low card. She bets only after player I has "advertised" a high card by betting high. Opportunity for subtlety on her side is therefore limited. But player I's go-for-broke behavior is much less easy to get some feeling for: and this is perhaps the lesson of the model. Intuition is *not* necessarily a good guide in these situations. Who would guess, before the event, that optimal behavior by player I includes betting high for certain with the worst possible hand? But this is characteristic, not only of player I's optimal behavior in this model, but of optimal play for straight two-person pot-limit poker in general. Cutler (1975) has shown, using a computer, that optimal play requires the opening player always to begin by betting the maximum on a really bad hand (like a ten-high bust). But, if dealt a somewhat better hand (like a pair of twos), he should call and fold if bet into.

Most poker players behave quite differently when playing in their parlors. They know they ought to bluff but they bluff too cautiously. In particular, they are reluctant to bluff on really bad hands. They figure that, if bluffing has to be done, let it be done on middle-range hands because there will then be some change of winning even if there is a showdown. Such thinking neglects the principal aim of bluffing, which is to make the opposition think it worthwhile to put money in the pot on those occasions when you do have a good hand. Von Neumann's second poker model indicates that this aim is best achieved by aggressive bluffing of the crudest kind.

NOTES

1 Even if the information on strategy spaces were incomplete, Harsanyi (1967–8) has described a device which transfers the uncertainty to the payoff functions.

2 It has become fashionable to write "asymmetric information" instead of "incomplete information." When this terminology is used, it must be kept in mind that a game of asymmetric information may well be symmetric in the mathematical sense.

3 Strategy 2 gives c at most 8 if she is playing a and at most -12 if playing b. The former occurs with probability 1/10 and the latter with probability 9/10. Her expected payoff from strategy 2 is therefore at most $[8+9(-12)/10$. A similar calculation shows that she gets at least $[-8+9(+8)]/10$ from strategy 1.

4 All information sets in this example are reached with *positive* probability, whatever strategies are used. In examples which have information sets which are reached with zero probability when equilibrium strategies are used, a more careful analysis is necessary.

5 When the opening chance move does not represent an actual event but, instead, is simply a mathematical device representing the beliefs of the players, then all the actors in each population (except one) are just ghosts with no real existence. Each ghost represents an actual player as that player "might have been" if he or she had enjoyed a different life experience. Does it make sense to analyze such a game as a contest? Where did the necessary common knowledge come from? Is it not possible that introspection might lead to implicit coordination between players and their ghostly alter egos? This point is perhaps recognized in a tendency to abandon Nash equilibrium in favor of Aumann's (1987) "correlated equilibrium." But such an abandonment raises further foundational difficulties.

6 This is not inconsistent with what was said about consistent beliefs in section 3.1. Suppose a two-card deck may consist of one of the three possibilities {H, H}, {H, L}, and {L, L} and that it is common knowledge that each member of a large population attaches probability 1/3 to each possibility. All members of the population are now separately shown one card of the deck chosen randomly and independently on each occasion. They then update their beliefs using Bayes' rule. Those who have seen the card H now attach probabilities 2/3, 1/3, and 0 respectively to the decks {H, H}, {H, L}, and {L, L}. Given this distribution, the probability that a randomly sampled card will be L is 1/6. Those who have been shown the card H will therefore believe that, with probability 1/6, an opponent will believe that the appropriate probabilities are 0, 1/3, 2/3 respectively for the decks {H, H}, {L, H}, and {L, L}.

7 With some tie-breaking rule in case several bids are all highest.

8 Concern over what happens if the hands are equal is unnecessary since this happens with zero probability.

9 Subsidiary conditions (that p and q are Lebesgue measurable and that sets of measure zero are to be neglected) are also necessary for strict accuracy.

REFERENCES

Aumann, R. 1976: "Agreeing to disagree," *Annals of Statistics* 4, 1236–9.

Aumann, R. 1987: "Correlated equilibrium as an expression of Bayesian rationality," *Econometrica* 55, 1–18.

Aumann, R., and Maschler, M. 1972: "Some thoughts on the minimax principle," *Management Science* 18, 54–63.

Binmore, K. 1987: "Modelling rational players I," *Economics and Philosophy* 3, 179–214.

Cutler, W. 1975: "An optimal strategy for pot-limit Poker," *American Mathematical Monthly* 82, 368–76.

De Finetti, B. 1974: *Theory of Probability.* New York: Wiley.

Gibbard, A. 1973: "Manipulation of voting schemes: a general result," *Econometrica* 41, 587–602.

Harsanyi, J. 1967–8: "Games of incomplete information played by Bayesian players, I, II and III," *Management Science* 14, 159–82, 320–4, 486–502.

Harsanyi, J. 1973: "Games with randomly distributed payoffs: a new rationale for mixed-strategy equilibrium points," *International Journal of Game Theory* 2, 235–50.

Luce, D., and Raiffa, H. 1957: *Games and Decisions.* New York: Wiley.

Myerson, R. 1979: "Incentive compatibility and the bargaining problem," *Econometrica* 47, 61–73.

Ramsey, F. 1931: "Truth and probability," in *Foundations of Mathematics and other Logical Essays*, ed. Braithwaite. London: Routledge and Kegan Paul.

Savage, L. 1954: *Foundations of Statistics.* New York: Wiley.

Von Neumann, J. and Morgenstern, O. 1944: *The Theory of Games and Economic Behavior.* Princeton, N.J.: Princeton University Press.

4 Common knowledge and game theory

(written jointly with Adam Brandenburger)

I know not what I know not.

St Augustine, *Confessions*

4.1 Introduction

It is traditional to introduce the subject of common knowledge with a little story. The following version is quoted from Littlewood's (1953) *Mathematical Miscellany*:

> Three ladies, A, B, and C in a railway carriage have dirty faces and are all laughing. It suddenly flashes on A: why doesn't B realize C is laughing at her? – Heavens! *I* must be laughable.

A more elaborate version of the story[1] concerns three less frivolous ladies called Alice, Bertha, and Cora. Each lady blushes if and only if she knows her own face to be dirty. All three have dirty faces but nobody blushes until a clergyman enters the carriage and remarks that there is a lady in the carriage with a dirty face. It is now impossible that nobody will blush.

If it were true that nobody blushes then Alice could reason as follows:

Alice: Suppose my face is clean so that Bertha and Cora can exclude me from the set of dirty-faced ladies. Then Bertha could argue as follows:

Bertha: Suppose my face is clean so that Cora can exclude me from the set of dirty-faced ladies. Then Cora could argue as follows:

Cora: The clergyman's statement tells me that the set of dirty-faced ladies is not empty. Neither Alice nor Bertha are in the set. Hence I am in the set and must blush.

Bertha: My simulation of Cora's reasoning informs me that she should

---

This chapter is an enlarged and much revised version of Binmore and Brandenburger (1988). We are grateful to Robert Aumann, Eddie Dekel, John Geanakoplos, Ariel Rubinstein, Dov Samet, and Avner Shaked for their comments and corrections.

blush if I have a clean face. Since she has not blushed, I have a dirty face and must blush myself.

Alice: My simulation of Bertha's reasoning informs me that she should blush if I have a clean face. Since she has not blushed, I have a dirty face and must blush myself.

But the argument began with the hypotheses that nobody blushes. Since this hypothesis has been contradicted, somebody must blush. The same reasoning, of course, applies however many ladies there might be.

Why did the ladies in the story have to wait for the clergyman's intervention before blushing? After all, he only told them that there was a dirty-faced lady in the carriage and each lady *already knew* that there were at least two dirty-faced ladies in the carriage. But, to carry the argument through, the ladies need also to know *what the other ladies would know* under various hypothetical circumstances. Thus Alice needs to know what Bertha would know if Alice had a clean face. Moreover, Alice needs to know what Bertha would know about what Cora would know if Alice and Bertha had clean faces. And so on through as many levels of knowing as there are ladies in the carriage. It is *this* information that is supplied by the clergyman's announcement.

The clergyman's announcement ensures that it will be *common knowledge* that a dirty-faced lady is present whenever there actually is a dirty-faced lady present. For an event to be common knowledge, the requirement is not only that every knows it, but that everybody knows that everybody knows it, and everybody knows that everybody knows it; and so on.

This is an abstruse-looking definition, but it is an important one. The reason is that any equilibrium notion that incorporates some measure of self-prophesying *necessarily* entails common knowledge requirements of some kind, although these are seldom stated explicitly. In equilibrium, agents optimize given their predictions of the future. The future therefore partly depends on the predictions the agents make. Where do these predictions come from? If they are well founded, they will be based on an agent's knowledge. In the case of only two agents, this means that agent 1 must know at least something about agent 2's knowledge, because the future depends partly on agent 2's predictions. But a relevant part of agent 2's knowledge will be what agent 2 knows about agent 1's knowledge. And so on.

The background intuition for such discussions is that rational agents cannot "agree to disagree" (Aumann, 1976). Suppose, for example, that two agents agree to disagree about the probability p that heads will result from tossing a weighted coin. Agent 1 insists that p is almost certainly 0.501 and agent 2 insists that p is almost certainly 0.499. The two agents then have the economist's equivalent of a philosopher's stone, which they can use to resolve all conflicts of interest between them without the need for compromise on either side. They can agree to resolve any issue entirely in favor of agent 1 if the number of heads appearing in 1,000,001 tosses exceeds the number of tails, and entirely in favor of agent 2 if not.

But, if both agents simultaneously proposed such a deal, would they go through with it? If both agents are rational and have reason to believe the other is rational, then there are grounds for supposing otherwise. After the deal has been proposed, each agent now knows something about what the other agent knows. This *new* information should lead both to revise their estimates of p.

The story of the dirty-faced ladies already embodies the essential point. Mutual observation of agents' behavior can lead to information becoming common knowledge that was not common knowledge before.

Perhaps the most important area in which such problems arise is in the study of rational expectations equilibria in the trading of risky securities. How can there be trade if everybody's willingness to trade means that everybody knows that everybody expects to be a winner (see Milgrom and Stokey, 1982, Geanakoplos, 1988)? Since risky securities are traded on the basis of private information, there must presumably be some "agreeing to disagree" in the real world. But to assess its extent and its implications, one needs to have a precise theory of the norm from which "agreeing to disagree" is seen as a deviation.

The beginnings of such a theory are presented here. Some formalism is necessary in such a presentation because the English language is not geared up to express the appropriate ideas compactly. Without some formalism, it is therefore very easy to get confused. However, nothing requiring any mathematical expertise is to be described.

4.2 Knowledge

The account begins with a set Ω of possible *states of the world*. To keep things simple Ω will always be assumed to be a *finite* set when formal matters are under discussion. A subset E of Ω is to be identified with a possible *event*.

To discuss what an individual i knows, his *knowledge operator* K is introduced. For each event E, the set KE is the set of states of the world in which individual i knows that E has occurred. Or, more briefly, KE is the event that i knows that E has occurred.

For example, in the story of the dirty-faced ladies, a state space Ω with eight states is required. The eight states of the world will be numbered as indicated in figure 4.1. The event that Alice's face is dirty is then $D = \{2, 5, 6, 8\}$. If blushing were not part of the story, she would *know* that her face was dirty after the clergyman's announcement only in state 2. Writing K_A for Alice's knowledge operator, it would then be true that $K_A D = \{2\}$.

States	1	2	3	4	5	6	7	8
A's face	Clean	Dirty	Clean	Clean	Dirty	Dirty	Clean	Dirty
B's face	Clean	Clean	Dirty	Clean	Dirty	Clean	Dirty	Dirty
C's face	Clean	Clean	Clean	Dirty	Clean	Dirty	Dirty	Dirty

Figure 4.1

What should be assumed about the knowledge operator? The properties usually considered are listed below. All but the final property (K4) will be taken for granted throughout the paper. Property (K4) is more controversial and its use will be postponed until section 4.5.

(K0) $K\Omega = \Omega$

(K1) $K(E \cap F) = KE \cap KF$

(K2) $KE \subseteq E$ (axiom of knowledge)

(K3) $KE \subseteq K^2E$ (axiom of transparency)

(K4) $(\sim K)^2 E \subseteq KE$ (axiom of wisdom)

The first two properties are book-keeping assumptions. Notice that it follows from (K1) that, if $E \subseteq F$, then $KE \subseteq KF$. Bacharach (1987) refers to (K2) as the "axiom of knowledge" on the grounds that it expresses the requirement that one can only really *know* that something has happened if it actually *has* happened. For similar reasons, Geanakoplos (1988) refers to "nondeluded" individuals in this connection. In (K3), K^2E stands for $K(KE)$ and hence is the event that the individual knows that he knows that E has occurred. Thus (K3) means that the individual cannot know that something has happened without knowing that he knows it. This explains Bacharach's (1987) use of the terminology "axiom of transparency" for (K3). Notice that (K2) and (K3) together imply that $KE = K^2E$. In (K4), $\sim KE$ stands for the complement of KE, and hence the condition requires that if the individual does *not* know that he does *not* know something, then he knows it. Discussion of this "axiom of wisdom" is postponed until it is used in section 4.4.

A *truism* T will be defined to be an event that cannot occur without the individual knowing that it has occurred. This translates into symbols as $T \subseteq KT$. For example, the event that a dirty-faced lady is in the railway carriage is a truism for Alice provided that the clergyman can be relied upon to draw her attention to the fact whenever it happens and never when it does not.

If one thinks of truisms as embodying the essence of what is involved in making a direct observation, then there is a sense in which all knowledge is derived from truisms. The following proposition expresses the idea formally.

PROPOSITION 4.1

An event E is known to have occurred ($\omega \in KE$) if and only if a truism T has occurred which implies E ($\omega \in T \subseteq E$).

The proposition has little content but will serve to illustrate the use of properties (K0) through (K3). Note first that, by (K2), the criterion for a triusm may be simplified to

$$T = KT$$

Since $KE = K^2 E$ for all events E, it follows that truisms are just the events of the form KE, where E is any subset of Ω.

Returning to proposition 4.1, observe that, if $\omega \in KE$, then $T = KE$ is a truism satisfying $\omega \in T \subseteq E$. On the other hand, if $\omega \in T \subseteq E$, then $KT \subseteq KE$ and so $\omega \in T = KT \subseteq KE$. Thus $\omega \in KE$.

4.3 Common knowledge

The formulation of common knowledge in terms of everybody knowing that everybody knows and so on was first given by Lewis (1969) in a philosophical study of conventions. Lewis attributes the basic idea to Schelling (1960). Aumann (1976) came up with the idea independently in a somewhat different context. His formulation is important because it provides a precise characterization of when an event is common knowledge that does not require thinking one's way through an infinite regress.

For the case of three dirty-faced ladies, the (everybody knows) operator is defined by

$$(\text{everybody knows})E = K_A E \cap K_B E \cap K_C E$$

The event that everybody knows that everybody knows E is then

$$(\text{everybody knows})^2 E$$

and the event that everybody knows that everybody knows that everybody knows E is

$$(\text{everybody knows})^3 E$$

With these preliminaries out of the way, it is now possible to define the event

$$(\text{everybody knows})^\infty E$$

to be the intersection of all sets of the form (everybody knows)$^n E$.

Lewis's (1969) criterion for common knowledge can now be expressed formally. An event E is common knowledge when ω occurs if and only if

$$\omega \in (\text{everybody knows})^{\infty} E$$

Some observations about the (everybody knows) operator will be helpful in getting to Aumann's (1976) criterion. It is being assumed that each individual's knowledge operator (K_A, K_B, and K_C in the case of the three dirty-faced ladies) satisfies (K0) through (K3). It follows that the operator $K = (\text{everybody knows})$ satisfies (K0) through (K2). (Usually (K3) will not be satisfied.) From (K2) one may conclude that the sets $(\text{everybody knows})^n E$ are shrinking in that each contains its predecessor. Since Ω is a finite set, a positive shrinkage can only occur a finite number of times. Thus, for some N,

$$(\text{everybody knows})^n E = (\text{everybody knows})^{\infty} E$$

for all $n \geqslant N$.

This point is made to facilitate checking that the *common* knowledge operator $K = (\text{everybody knows})^{\infty}$ satisfies not only (K0) through (K2) but (K3) as well. The following analog to proposition 4.1 must therefore be true.

PROPOSITION 4.2

An event E is common knowledge ($\omega \in (\text{everybody knows})^{\infty} E$) if and only if a common truism T has occurred which implies E ($\omega \in T \subseteq E$).

A common truism is, of course, an event T that satisfies $T = (\text{everybody knows})^{\infty} T$. However, the infinite regress in this definition can be eliminated. The criterion

$$T = (\text{everybody knows}) T$$

is equivalent[2] and simpler. This in turn can be replaced by the even simpler criterion of the following proposition.

PROPOSITION 4.3

An event T is a common truism if and only if it is a truism for each individual separately.

A proof will be given for the case of two dirty-faced ladies. If T is a truism for eady lady separately, then $T = K_A T$ and $T = K_B T$. Thus $T = T \cap T = K_A T \cap K_B T = (\text{everybody knows}) T$. If T is a common truism, then $T = K_A T \cap K_B T \subseteq K_A T \subseteq T$ (by K2). Hence $T = K_A T$ and so T is a truism for Alice and a similar argument shows T to be a truism for Bertha.

(Mathematicians may prefer to express proposition 4.3 in terms of the set $\mathscr{T}$ of common truisms and the sets $\mathscr{T}_A$, $\mathscr{T}_B$, and $\mathscr{T}_C$ of individual truisms. It then asserts that $\mathscr{T} = \mathscr{T}_A \cap \mathscr{T}_B \cap \mathscr{T}_C$. One may add that all these sets are topologies, being closed under intersections and unions (recall that Ω is finite). Thus $\mathscr{T}$ is the finest common coarsening of the topologies $\mathscr{T}_A$, $\mathscr{T}_B$, and $\mathscr{T}_C$.)

The importance of what have been called common triusms in the preceding discussion has been emphasized by a number of authors. Mondered and Samet (1988) call them "evidently known events." Geanakoplos (1988) prefers "necessarily known events." Milgrom (1981) speaks of "public events."

4.4 Possibility

After ω occurs, an individual will know that certain states are impossible. For example, if T is any truism containing ω, then he will know that states outside T are impossible. Thus the set $P(\omega)$ of *possible* states when ω occurs lies inside each truism T containing ω. But $P(\omega)$ must itself be a truism because the individual cannot evade knowing what he regards as possible. Thus $P(\omega)$ is the *smallest*[3] truism containing ω.

Examples of possibility sets can be found by returning to the story of the dirty-faced ladies. The state space Ω is described in figure 4.1. Figure 4.2. shows the possibility sets for each lady *before* the clergyman provides any information. For example, whatever Alice learns about the faces of the other ladies, it remains possible for Alice that her own face is either clean or dirty. Thus $P_A(1) = P_A(2) = \{1, 2\}$.

Figure 4.3 shows the possibility sets, *after* the clergyman's announcement but *before* any blushing takes place, on the assumption that all are aware that the clergyman *invariably and reliably* comments on the presence of a dirty face. When Alice sees two clean faces, she can now deduce the state of her own face from whether or not the clergyman makes an announcement. Thus $P_A(1) = \{1\}$ and $P_A(2) = \{2\}$.

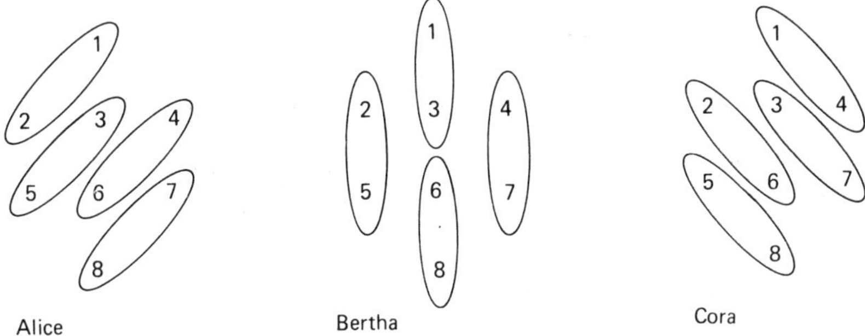

Figure 4.2

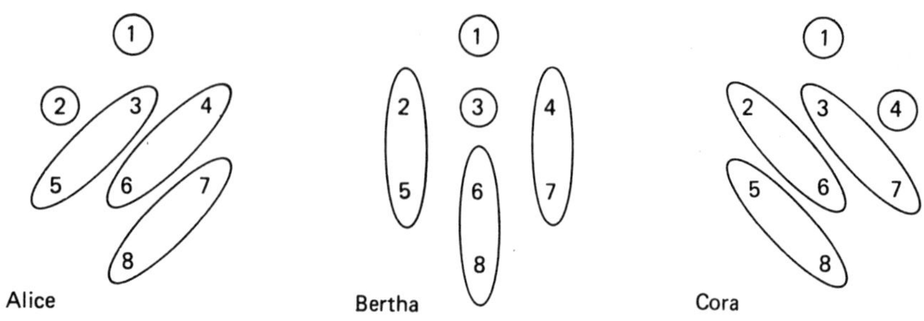

Figure 4.3

A more tricky example is obtained by having the clergyman announce that a dirty-faced lady is present if and only if *two* or more dirty-faced ladies are actually present. The ladies know how he will behave *except* in the case when he sees precisely *two* dirty-faced ladies in which case they are *unsure* whether or not he will make an announcement. Figure 4.4 illustrates the possibility sets.

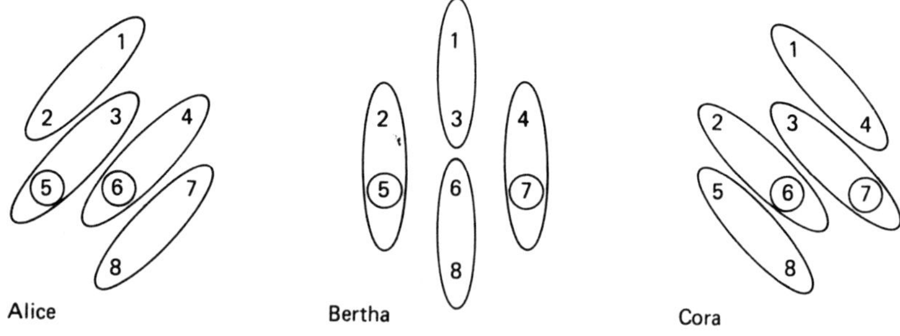

Figure 4.4

In the case when Bertha's face is dirty and Cora's is clean, Alice can deduce from the fact that an announcement is made that her face is dirty. Moreover, the clergyman actually will announce that there is a dirty face when Alice's face is dirty. Thus $P_A(5) = \{5\}$. But $P_A(3) = \{3, 5\}$ because, if 3 occurs, then the clergyman will make no announcement and, from this eventuality, Alice can deduce nothing.

Because $P(\omega)$ is the *smallest* truism containing ω, proposition 1 can be rephrased as follows.

PROPOSITION 4.4

When ω occurs, E is known if and only if $P(\omega) \subseteq E$ (i.e. everything possible implies E).

Proposition 4.2 can be similarly rephrased. In its rephrased form it is Aumann's (1976) criterion for an event to be common knowledge. Some preliminary explanation is necessary to make this point.

The smallest *common* truism containing ω will be denoted $M(\omega)$. One may think of this as the set of states deemed to be possible by the community as a whole. It includes not only each state that somebody thinks to be possible, but also each state that somebody thinks somebody else thinks to be possible, and so on.

To find $M(\omega)$, one may use proposition 4.3 and the fact that truisms are simply unions[4] of possibility sets. To find $M(2)$ in figure 4.3, for example, look for the smallest set containing 2 which is simultaneously the the union of possibility sets belonging to each of Alice, Bertha, and Cora separately. Since $M(2)$ contains 2, it must contain 5 because of Bertha. Hence it must contain 8 because of Cora, and therefore 7 because of Alice. Proceeding in this way one finds that $M(2) = \{2, 3, 4, 5, 6, 7, 8\}$.

Figure 4.5 illustrates the communal possibility sets for each of the situations illustrated in figures 4.2, 4.3, and 4.4. (Mathematicians may choose to call M the meet of P_A, P_B, and P_C by analogy with the usage for partitions.)

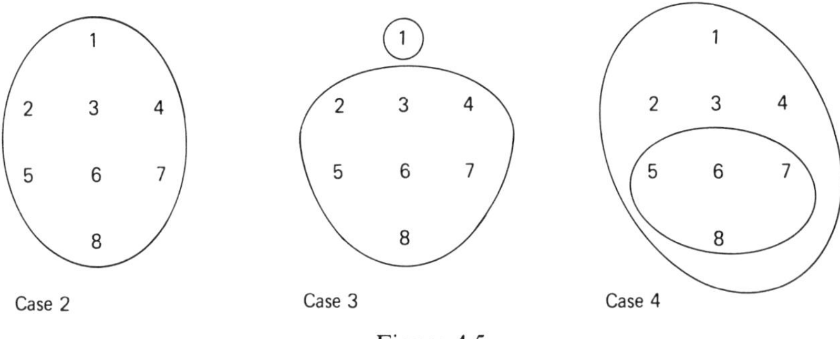

Figure 4.5

PROPOSITION 4.5 (AUMANN, 1976)

When ω occurs, E is common knowledge if and only if $M(\omega) \subseteq E$ (i.e. everything communally possible implies E).

It follows, for example, that in the information set-up described by figure 4.3, no matter what happens, the set $D = \{2, 5, 6, 8\}$, which represents the event that Alice's fact is dirty, never becomes common knowledge.

In the story of the dirty-faced ladies given in section 4.1, more information was provided than is summarized in figure 4.3. It was also given that ladies blush when they know their faces to be dirty. Figures 4.6, 4.7, and 4.8 give three different knowledge configurations that are consistent with the story.

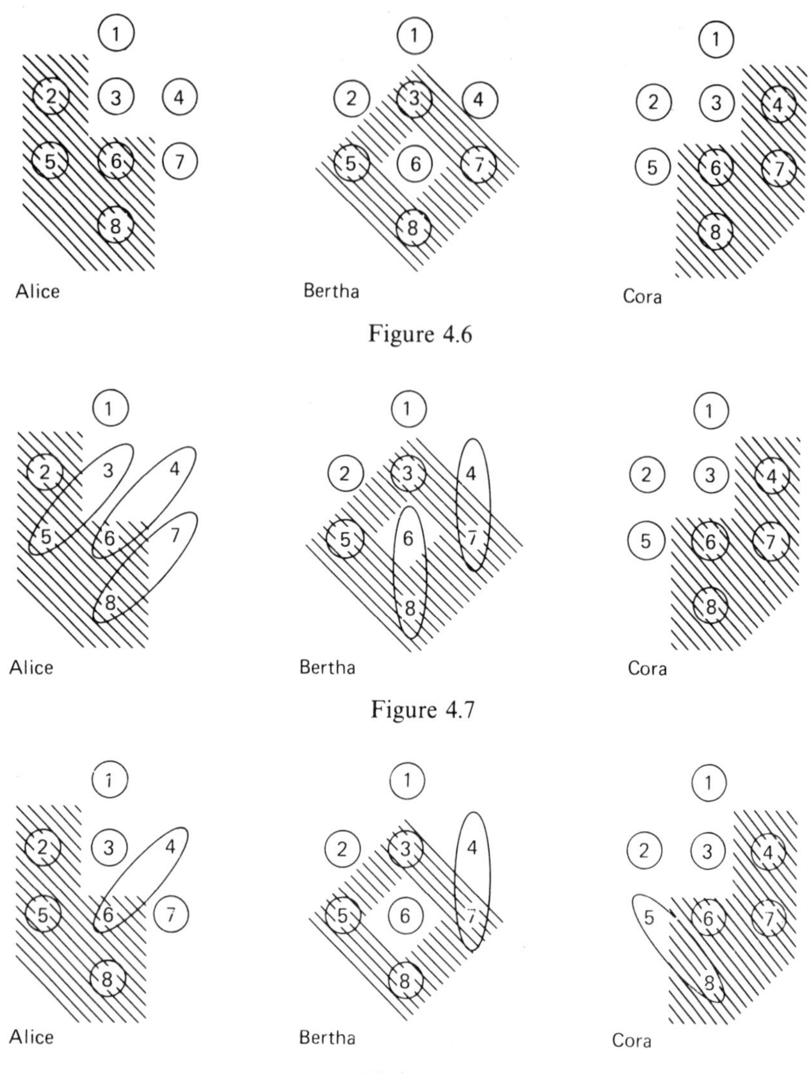

Figure 4.6

Figure 4.7

Figure 4.8

The shaded regions indicate states in which a lady has a dirty face. If each state in a possibility set is shaded, a lady will blush if one of these state occurs. Otherwise she will not blush.

Consider figure 4.7 by way of example. It has to be checked that the configuration is consistent with the data given in the story. Suppose, for instance, that state 6 occurs. Cora will then blush, but not Alice or Bertha. But does the fact that Cora blushes not convey information to Alice and Bertha? Notice that $P_A(6) = \{4, 6\}$. Thus, if 6 occurs, Alice knows that *either* 4 *or* 6 have occurred. But Cora would blush in *both* cases. Hence the fact that Cora blushes does not tell Alice anything. Nor does the fact that Bertha does *not* blush. She would not blush either in case 4 or in case 6.

In the story, state 8 is what actually occurs. Thus, for the configuration in figure 4.7, Cora blushes but Alice and Bertha do not. Observe that $M(8) = \{4, 6, 7, 8\}$, which is precisely the event that Cora has a dirty face. This becomes common knowledge whenever it occurs. (It is, of course, the smallest common truism containing 8.)

A final lesson can be extracted from the dirty-faced ladies. It concerns the question of *how* things come to be known. In the story, no information is offered on the *process* by means of which the ladies learn. A conclusion is reached by indirect reasoning. For example, figure 4.3 is inconsistent with what is given about blushing behavior because, if 2 occurs, the Alice will blush, whereas, if 5 occurs, she will not. Thus Bertha can distinguish between 2 and 5 and so it is incorrect to assert that $P_B(2) = P_B(5) = \{2, 5\}$. Only certain knowledge configurations are consistent with the data of the story. Those illustrated in figures 4.6, 4.7, and 4.8 are examples. The argument given in section 4.1 shows that, for each such example, at least one of the ladies will blush when 8 occurs. (In fact, a lady will blush whenever a dirty-fased lady is actually present.)

How might figure 4.7 arise? One "explanation" postulates that the opportunity to blush rotates among the ladies, starting with Alice. The knowledge situation then evolves as follows.

1. The ladies observe the faces of their companions. This leads to figure 4.2 in which $M(8) = \Omega$.
2. The clergyman announces the presence of a dirty face whenever a dirty-faced lady is present. This leads to figure 4.3 in which $M(8) = \{2, 3, 4, 5, 6, 7, 8\}$.
3. Alice has the opportunity to blush (but not Bertha or Cora). This leads to figure 4.9 in which $M(8) = \{3, 4, 5, 6, 7, 8\}$.

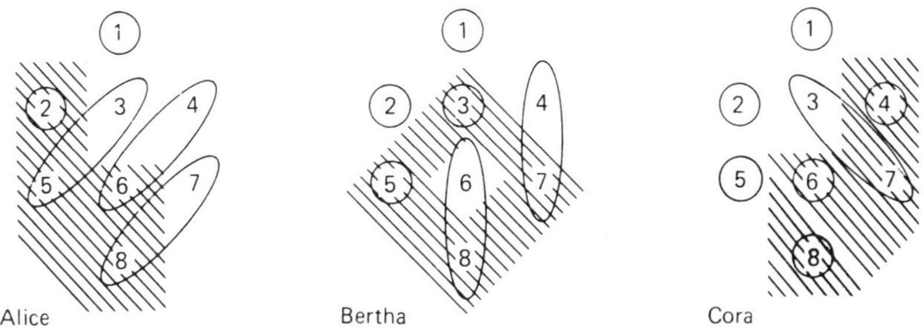

Figure 4.9

4 Bertha has the opportunity to blush (but not Cora). This leads to figure 4.7 in which $M(8)=\{4, 6, 7, 8\}$.

5 Cora has the opportunity to blush and does so when 8 occurs. But figure 4.7 remains unchanged and continues unchanged even if further rounds of blushing are introduced.

An "explanation" of figure 4.6 can be obtained by supposing that all three ladies have the opportunity to blush precisely one second after the clergyman's announcement, and then again precisely two seconds after, and so on. Figure 4.8 can be "explained" by having Alice and Bertha blush simultaneously in rotation with Cora.

More on the question of how knowledge evolves over time will appear in section 4.12.

4.5 Information partitions

In the previous section, the idea of a possibility set was developed taking the knowledge operator K as a primitive. When K satisfies (K0) through (K3), possibility sets necessarily satisfy

(P2) $\quad \omega \in P(\omega)$

(P3) $\quad \zeta \in P(\omega) \Rightarrow P(\zeta) \subseteq P(\omega)$

for all ζ and ω. The first of these simply says that the state that actually occurs is always regarded as possible. The second is a little more complicated. It says that, if something is possible, then anything that would be regarded as possible in that state must also be regarded as possible in the current state.

Section 4.4 makes it clear that, for practical purposes the possibility operator P is much more convenient to work with than the knowledge operator K. One can, in fact, pass back and forward between the two without difficulty since (K0) through (K3) hold if and only if (P2) and (P3) hold. If one begins with P, the operator K may be defined by

$$KE = \{\omega\colon P(\omega) \subseteq E\}$$

In applications, it is usually taken for granted that the possibility operator *partitions* the state space. This means that distinct possibility sets have no points in common. Figures 4.2 and 4.3 illustrate six possible partitions on the state space $\Omega = \{1, 2, 3, 4, 5, 6, 7, 8\}$. Figure 4.4 illustrates three situations in which Ω is not partitioned. (Notice that $P_A(5) = \{5\}$ and $P_A(3) = \{3, 5\}$. Thus $P_A(5)$ and $P_A(3)$ are distinct sets but they have the point 5 in common.)

In game theory, possibility sets which partition the state space are called *information sets*, following Von Neumann and Morgenstern. Figure 4.10

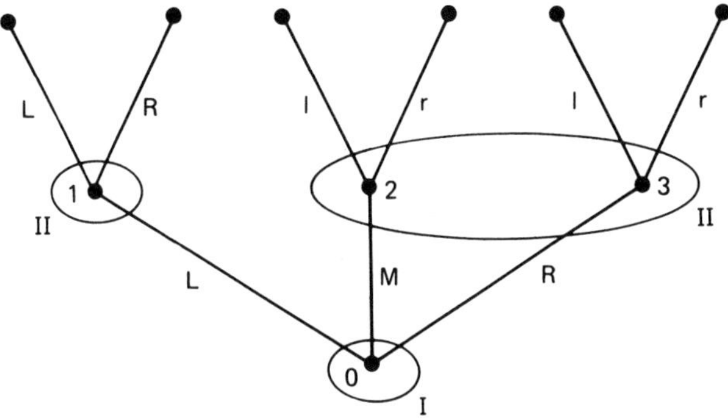

Figure 4.10

illustrates a simple game tree. When player II gets to move, her state space is $\Omega = \{1, 2, 3\}$. The information set enclosing 2 and 3 indicates that, if player I chooses M or R, then player II will not know which of these was chosen.

In formal terms, the requirement that the possibility sets partition Ω may be expressed as

$$\text{(P4)} \qquad \zeta \in P(\omega) \Rightarrow P(\zeta) = P(\omega)$$

for all ζ and ω. Conditions (P2) through (P4) are equivalent to (K0) through (K4). Thus (K4), the controversial axiom of wisdom, appears on the scene. The implications are discussed in section 4.6. This section closes with a comment on the common knowledge operator.

If P_A, P_B, and P_C partition Ω, then so does their meet M. Consider, for example, cases 2 and 3 in figure 4.5 which illustrates the meets of the partitions of figures 4.2 and 4.3. A consequence is that, if each individual's knowledge operator satisfies (K0) through (K4), then so does the common knowledge operator.

4.6 Small worlds

There is something paradoxical in the preceding section. The use of information partitions in a situation like the game of figure 4.10 seems more than harmless: it seems inevitable. On the other hand, using information partitions is equivalent to incorporating (K4) into the system. But do we really want to claim that, whenever we don't know that we don't know something, then we know it?

Let us return to Alice in figure 4.4. The story that goes with this will be told again. The clergyman may be in a good mood or a bad mood. If in a good mood he announces the presence of a dirty-faced lady if and only if all three ladies have dirty faces. If in a bad mood, he announces the presence of a dirty-faced lady if and only if at least two ladies have dirty faces. In figure 4.4, the clergyman is in a bad mood but Alice thinks he may be in either a good mood or a bad mood. The result is a system of possibility sets that do not partition Ω.

One reaction is that the state space Ω has been improperly specified since it pays no attention to the mood of the clergyman although this is clearly relevant. If 1, 2, 3, ..., 8 are as previously, but with the extra understanding that the clergyman is in a good mood, while I, II, III, ..., VIII are the corresponding states in which the clergyman's mood is bad, then Alice's knowledge configuration in figure 4.4 may be replaced by that of figure 4.11.

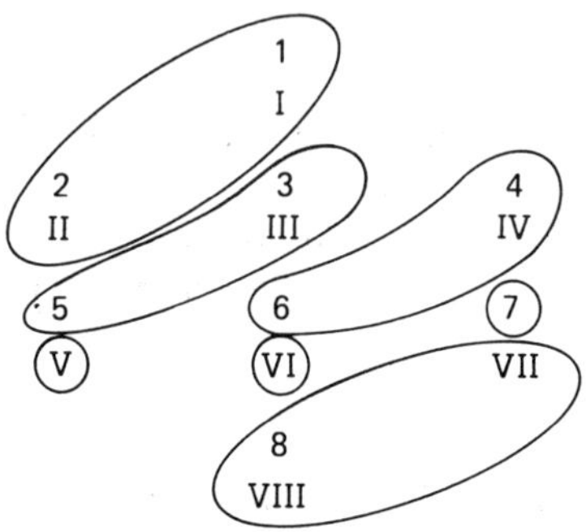

Figure 4.11

The result is an *information partition* of a *new* state space $\Omega^* = \{1, 2, \ldots, 8, \mathrm{I}, \mathrm{II}, \ldots, \mathrm{VIII}\}$. One can therefore regard the failure to arrive at an information partition in figure 4.4 as a consequence of a wrong modeling judgment about the relevant state space. Something which matters, namely the mood of the clergyman, has been omitted from the description of a state. Once this omission has been rectified, by constructing a new state space Ω^* consisting of more fully described states, the problem disappears.

For this kind of reason, Bayesian decision theorists often urge that care be taken to begin with a state space Ω in which the description of a state is *all-inclusive*. This is sound advice if everything that is to be included can be tagged and enumerated at the outset.

To see why this proviso is attached, consider the following argument in defense of information partitions. Recall that the formal requirement for an information partition is (P4) of section 4.5.

Suppose that an individual finds himself holding the view that only states in $P(\omega)$ are possible. If $P(\zeta) \neq P(\omega)$, he can then deduce that it is impossible that ζ has occurred, because, if ζ had occurred, then his current view of what is possible would differ from the view he actually has. Thus $P(\zeta) \neq P(\omega)$ implies that $\zeta \notin P(\omega)$ or, what is the same thing, $\zeta \in P(\omega)$ implies $P(\zeta) = P(\omega)$.

This is the requirement (P4) for an information partition. Now tell the same story but with "possible" replaced everywhere by "conceivable."

From $P(\zeta) \neq P(\omega)$, the conclusion that ζ is inconceivable should follow. But if ζ is truly inconceivable, how does the individual manage to "know" the set $P(\zeta)$ of those states he would regard as conceivable if the inconceivable state ζ were to occur? If the individual is assumed only to be able to explore the implications of conceivable states, then only (P3) can be justified. The latter only requires that anything that is conceivable at a conceivable state be conceivable.

Samet (1987) observes that, without information partitions, an individual might be "ignoring his ignorance" to some extent. To see this point, take complements on both sides of (K4) to obtain $\sim KE \subseteq K(\sim KE)$. This says that, if an individual is ignorant of something, then he *knows* he is ignorant of it. If information partitions fail to exist because (K4) is not satisfied, it follows that there is at least one event of which the individual is ignorant but does not know it. The argument is then that, since he could examine the contents of his own mind and enumerate those events that he knows and those that he does not know, if he does not know that he is ignorant of an event, then it is because he ignores his ignorance. This seems thoroughly reprehensible. On the other hand, who could be blamed for being unaware of being unaware of something?

Two versions of the same story have been told to make it clear the English language has no difficulty in distinguishing between "closed universe" and "open universe" situations. By a closed universe is meant one in which all the possibilities can be exhaustively enumerated in advance, and all the implications of all possibilities explored in detail so that they can be neatly labeled and placed in their proper pigeonholes. In discussing the foundations of Bayesian decision theory, Savage (1954) uses the term "small world" for such a state space. Statisticians' sample spaces are invariably small worlds in this sense, with the set $\Omega = \{1, 2, 3, 4, 5, 6\}$ of possible ways a die can fall serving as the archetypal example. Within a small world, the arguments in favor of (K0) through (K4), and of standard Bayesian principles, are very powerful (although, as the example at the beginning of this section shows, a bad modeling judgment can generate a world which is *too* small!)

However, there are difficulties with working in a small world. Certain things can only be expressed *informally*. For example, in game theory, it is typically understood that the structure of the game tree is to be common knowledge. But there is no way of expressing this within a formalism that takes as its small world the set $\Omega = \{0, 1, 2, 3\}$ of decision nodes in figure 4.10. Just as the use of information partitions has been defended by appealing to unformalized understandings about what an individual should be expected to know as a consequence of examining the workings of his own mind, so the understandings about what is informally regarded as common knowledge will appear on the scene only when the type of game-theoretic analysis employed comes under attack.

Of course, the larger the small world, the more that can be expressed up-front in formal terms. Papers that explore what can be done in this direction are those by Aumann (1987), Bacharach (1985, 1987), Brandenburger and Dekel (1985), Kaneko (1987), Mertens and Zamir (1985), Samet (1987), Shin (1987), and Tan and Werlang (1985). There is also a large and relevant philosophical literature on the theory of knowledge and epistemic logic (see Halpern, 1986, and the references therein).

The deep issues discussed in these papers certainly need to be thoroughly explored, but perhaps some healthy skepticism is appropriate about how useful their conclusions are likely to prove in the foundations of game theory. Any formal characterization of how we acquire knowledge is bound to be an oversimplification and hence will generate distortions if pushed beyond its limitations. In particular, one has to expect distortions if "closed universe" methodologies are applied to "open universe" problems. This risk is greatest when attempts are made to interpret a state as incorporating a specification of the universe that is totally all-embracing. To know a state then includes knowing, not only everything there is to know about the state of the physical world, but also everything there is to know about everybody's *state of mind*, including their knowledge and beliefs.

The self-reference implicit in such an interpretation brings Gödel's theorem to mind. Recall that this says that any sufficiently complex formal deductive system cannot be complete unless it is inconsistent. That is to say, in the world of theorem-proving, the "open universe" is a *necessary* fact of life with which one has to learn to live. One is therefore perhaps entitled to be suspicious of theories of knowledge in which this fact of life is somehow evaded.

4.7 Algorithmic knowledge

The previous paragraph draws an analogy between proving a theorem and acquiring a piece of knowledge. This section briefly explores this idea a little further. The point is that, if what one knows is known by virtue of its being the result of applying a properly defined procedure or algorithm, then a very much more critical eye needs to be cast on assumptions (K0) through (K4) if the small-world approach advocated in the previous section is not to be followed. The notion that questions are to be settled algorithmically is captured by requiring that there exist a computer programmed for this purpose. The type of computer used in the argument is called a Turing machine. One may think of this as a machine with a *finite* program but with an indefinitely large storage capacity.

In preceding sections, doubts have been expressed about (K4) – the axiom of wisdom. The purpose of what follows is to direct suspicion at the very much more fundamental (K2) – the axiom of knowledge. This says that one cannot know an event that has not occurred. The corresponding condition for possibility sets is (P2), i.e. $\omega \in P(\omega)$. This means that any state that actually occurs cannot be regarded as impossible.

Suppose that, in each state ω, possibility questions are resolved by a Turing machine $S = S(\omega)$ that sometimes answers NO to properly coded questions of the form, "Is it possible that ... ?" If it answers otherwise, or not at all, possibility is conceded. (Timing issues are ignored.) Consider some specific question concerning the Turing machine N. Let the coded form of this question be $[N]$. Consider next the question, "Is it possible that Turing machine M would output NO when offered the input $[M]$?" Let the coded form of this question be $\{M\}$. Let T be a machine that, when offered $[M]$ outputs $\{M\}$. Then machine $R = ST$, constructed by composing S and T, responds to $[M]$ as S responds to $\{M\}$.

Suppose that R responds to $[R]$ with NO. Then S reports that it is impossible that $[R]$ responds to $[R]$ with NO. Thus $\omega \notin P(\omega)$. If it is to be denied that this can occur for any $\omega \in \Omega$, then it must be that R does *not* respond to $[R]$ with NO in *any* state. Nevertheless, S *always* reports that it is possible that R will respond to $[R]$ with NO. But then S will fail to report a piece of available knowledge.

Such an argument (which is a version of the halting problem for Turing machines loosely adapted from Binmore, 1984) can only serve to raise questions about how meaningful it is to work in terms of all-embracing descriptions of the universe and superhumanly endowed individuals who process such data. This goes also for more formal attempts in the same direction, such as that of Shin (1987). But perhaps it will be enough to explain why this paper persists with a small-world view in spite of the attractions of more ambitious foundational positions.

4.8 Agreeing to disagree?

In specifying who knows what during the course of a game, it is usual to take (K0) through (K4) for granted and therefore to work with information partitions. This practice will be followed throughout the chapter (except where otherwise stated), since the game trees with which we shall be concerned, like that considered already in figure 4.10, constitute very small worlds indeed.

If everybody always knows what has happened previously in the game, as in chess, then the game is one of *perfect information*. In such a game, each information set encloses only a single decision node. Poker, on the other hand, is a game of *imperfect information.* The players do not know what other players have been dealt when they make their bets.

A game of perfect information must be distinguished from a game of *complete information.* All games of perfect information are games of complete information, but so are all games of imperfect information. The information that is complete in a game of complete information is the information about the *structure of the game.* This can be classified under two headings: information about the *rules of the game* and information about the tastes and beliefs of the *players.* The former includes what the game tree is, where the information sets are placed, and who makes what decisions. The latter subsumes the players' Von Neumann and Morgenstern utilities for the possible outcomes of the game (i.e. the terminal nodes of the game tree) and the probabilities that the players attach to the various chance events that may occur in the game which lie outside the control of the players (as when the cards are shuffled and dealt in poker). All this information is taken to be *common knowledge* among the players in a game of complete information. (Since everybody is then necessarily in the *same* informational state before any moves are made, a game of complete information is also commonly referred to as a "game of symmetric information." However, this should not be taken to imply that games of incomplete information cannot have a symmetric structure.)

Games of *incomplete information,* in so far as they can be dealt with realistically at all, are studied by reducing them to games of complete information using an ingenious methodology introduced by Harsanyi (1967–8). For example, I may be in doubt whether my opponent at chess is really aiming to win or whether he actually wants to lose. Harsanyi would say that the game should then not be seen as a two-player game of perfect information, but as a game of imperfect information with at least three players in which the opening event is a *chance move* that selects my opponent from a menu of possible opponents with various different preferences.

A standard procedure is to take for granted that everybody attaches the *same* probabilities to the possible outcomes of such a chance move. This does not need much justifying when the chance move is that of shuffling and dealing at poker. But matters are less clear-cut with a chance move like that in the story attributed above to Harsanyi.

Recall that, in a game of incomplete information, the probabilities that the players attach to the possible outcomes of chance events are *common knowledge.* Aumann (1976) asked whether rational players can "agree to disagree" under such circumstances by maintaining different probabilities for the same event. His answer is that this is impossible. Intuitively, each player will use his knowledge of the other players' estimates of the probabilities to refine his own estimate and this will continue until all the estimates are the same. (See Bacharach, 1985, for a more general expression of the idea.)

However, Aumann requires a strong assumption that will be important throughout the rest of the paper. He likes to refer to this as the "Harsanyi doctrine" (see section 4.13). The idea is that the players' informational status can be modeled as follows. Before the rational players to be considered received any data at all, they were all in the *same* position and therefore assigned the *same* probabilities $\text{prob}(\omega) > 0$ to the states $\omega \in \Omega$. Moreover, these *prior* probabilities are common knowledge among the players. The Harsanyi doctrine is therefore that there is common knowledge of a common prior. Later, the players had different experiences which led them to revise their probabilities. The current probabilities they attach to the states $\omega \in \Omega$ are therefore posterior probabilities obtained by Bayesian updating from the given common prior. It is in this sense that the posterior probabilities q_A, q_B, and q_C for the event F attributed to Alice, Bertha, and Cora in the proof of the following proposition are to be understood.

PROPOSITION 4.6 (AUMANN, 1976)

If each agent's posterior probability for an event F is common knowledge when $\omega_0 \in \Omega$ occurs, then it is common knowledge that these probabilities are equal.

To see why this is true, focus on Alice. If ω_0 occurs, her posterior probability for F is q_A. The event that Alice actually observes when ω_0 occurs is $P_A(\omega_0)$. Thus

$$q_A = \text{prob}[F \mid P_A(\omega_0)]$$

It follows from proposition 4.3 that $M(\omega_0)$ is a union of a collection of Alice's information sets. Let these information sets be $Q_1, Q_2, \ldots, Q_k$. One of these sets is, of course $P_A(\omega_0)$.

Observe that

$$q_A = \text{prob}(F \mid Q_j)$$

for each $j = 1, 2, \ldots, k$. The reason[5] is that, if $\text{prob}(F \mid Q_j)$ were not equal to q_A then, as soon as q_A became common knowledge, it would become common knowledge that Q_j had *not* happened, and hence the event $\sim Q_j$ would be common knowledge. But proposition 4.5 then implies that $M(\omega_0) \subseteq \sim Q_j$, and so $Q_j \subseteq Q_1 \cup Q_2 \cup \ldots \cup Q_k = M(\omega_0) \subseteq \sim Q_j$, which is a contradiction.

It then only remains to observe that

$$\begin{aligned}\text{prob}[M(\omega_0)\cap F] &= \text{prob}(Q_1\cap F)+\ \ldots\ +\text{prob}(Q_k\cap F) \qquad (4.1)\\ &= \text{prob}(F|Q_1)\text{prob}(Q_1)+\ \ldots\ +\text{prob}(F|Q_k)\text{prob}(Q_k)\\ &= q_A(\text{prob}Q_1+\ \ldots\ +\text{prob}Q_k)\\ &= q_A\ \text{prob}\ M(\omega_0)\end{aligned}$$

Thus, since prob $M(\omega_0)>0$, $q_A=\text{prob}[F|M(\omega_0)]$. But the same is true for q_B and q_C and therefore

$$q_A=q_B=q_C$$

as required.

Generalizations of this result appear in a number of papers, notably those by Bacharach (1985), Geanakoplos (1988), and Rubinstein and Wolinsky (1988). In the latter paper, the authors consider general predicates G defined on subsets of Ω. For each $E\subseteq\Omega$, $G(E)$ takes one of the values TRUE or FALSE. If ω occurs, an individual will then say that G is true if and only if $G[(P(\omega)]=\text{TRUE}$. For example, with the given event F of proposition 4.6, one may define a predicate $G_{F,q}$ by taking $G_{F,q}$ to be TRUE if and only if $\text{prob}(F|E)=q$. When Alice announces that her probability for F is q_A, she is then expressing her faith in the statement $G_{F,q_A}[P_A(\omega)]=\text{TRUE}$.

In obtaining a generalization of proposition 4.6, Rubinstein and Wolinsky point out that the essential feature of the predicates G_A, G_B, and G_C under discussion is that they be preserved under unions. That is to say, for any disjoint events S and T, if $G_i(S)=G_i(T)=\text{TRUE}$, then $G_i(S\cup T)=\text{TRUE}$. With this hypothesis, they prove that, if it is common knowledge in state ω_0 that each agent i holds that predicate G_i is true, then there is at least one event $E_0\neq\varnothing$ for which $G_i(E_0)$ is true for all i.

The proof consists simply of the observation that one may take $E_0=M(\omega_0)$ because

$$M(\omega_0)\subseteq\{\omega\colon G_i[P_i(\omega)]=\text{TRUE for all } i\}$$

With the notation of proposition 4.6, it then follows that

$$G_A(E_0)=G_A[M(\omega_0)]=G_A(Q_1\cup Q_2\cup\ \ldots\ \cup Q_k)=\text{TRUE}$$

because G_A is preserved under unions and $G_A(Q_j)=\text{TRUE}$ for each $j=1, 2, \ldots, k$. Proposition 4.6 follows immediately because $G_{F,q_i}(E)$ cannot be true for all i unless $q_A=q_B=q_C$.

Rubinstein and Wolinsky go on to discuss how other results of a similar nature can be derived. In particular, they obtain a result of Milgrom and Stokey (1982) on contingent contracts. Suppose that c is a contract signed by Alice, Bertha, and Cora before $\omega\in\Omega$ is realized. The contact is assumed to be *ex ante* efficient. This means that there is no alternative contract d with the property that

$$\mathscr{E}u_i[d(\omega), \omega] > \mathscr{E}u_i[c(\omega), \omega] \qquad (i = \text{A, B, C})$$

State ω_0 now occurs and each agent i is privately informed of $P_i(\omega_0)$. Is it possible that the agents will now be willing to renegotiate the contract at this interim stage? Or, to be more precise, can it be common knowledge that there is a contract e that each agent now prefers to c?

Milgrom and Stokey answer this question in the negative. Rubinstein and Wolinsky prove the result using the predicate $G_{u,e,c}$ which is defined to be true at E if and only if

$$\mathscr{E}\{u_i[e(\omega), \omega] | E\} > \mathscr{E}\{u_i[c(\omega), \omega] | E\}$$

This is preserved under unions and so, with the hypotheses of Milgrom and Stokey's question, a non-null event E_0 exists for which $G_{u_i,e,c}(E_0) = \text{TRUE}$ for each $i = \text{A, B, C}$. But this conclusion is inconsistent with the fact that c is *ex ante* efficient, because the contract d which is equal to e for $\omega \in E$ and equal to c for $\omega \notin E$ would then be an *ex ante* Pareto improvement on c. It follows that the contract e of Milgrom and Stokey's question cannot exist.

Returning to proposition 4.6, Samet (1987) has pointed out that this does not depend on having information partitions. The result survives without (K4). It is easy to see why. In the proof of proposition 4.6, it is only at step (4.1) that information partitioning is required. For this step to be valid, the possibility sets $Q_1, Q_2, \ldots, Q_k$ must not overlap. Without (K4), the sets may overlap. In the case $k = 2$, for example, one must then replace (4.1) by

$$\text{prob}[M(\omega_0) \cap F] = \text{prob}(Q_1 \cap F) + \text{prob}(Q_2 \cap F) - \text{prob}(Q_1 \cap Q_2 \cap F)$$

All is then well provided $Q_1 \cap Q_2$ can be expressed as the union of possibility sets. But this is guaranteed because the intersection of two truisms is again a truism. Thus, in Samet's words, we cannot "agree to disagree" even if we "ignore our ignorance."

Rubinstein and Wolinsky (1988) show that Samet's result can be generalized to predicates G that are preserved under differences. This means that if $G(S) = G(T) = \text{TRUE}$ and $S \subseteq T$, then $G(S \setminus T) = \text{TRUE}$. This criterion is satisfied by the predicate $G_{F,q}$ considered in connection with proposition 4.6, but not by the predicate $G_{u,e,c}$ considered when the Milgrom and Stokey result was discussed.

However, a pause for thought is perhaps appropriate. Returning again to proposition 4.6, one may ask whether, in a world in which we "ignore our ignorance," it really makes sense to proceed as though probabilistic assertions can sensibly be formulated and then manipulated according to the usual rules. Is it not a background presumption of probabilistic decision theories that information sets partition the universe? This is a question which returns us to the small-world issues of section 4.6 and anticipates some of what is to be said in section 4.13.

Any doubts that may arise seem amply justified, as the following example is intended to indicate. Return to Alice's informational configuration of figure 4.4, reproduced as the "Arabic world" on the left in figure 4.12. Recall that the clergyman announces that a dirty-faced lady is present if and only if two or more such ladies actually are present, but Alice only knows for sure what he will do when the actual number of dirty-faced ladies differs from two. Suppose that Alice attaches equal prior probabilities to each of the states in Ω. What probability should she attach to the event E that her own face is dirty, given that she sees that Bertha is dirty and Cora is clean? If this is posed as the question, "What is prob($\{3\}|\{3, 5\}$)?" then a conventional conditional probability calculation yields the answer $1/2$. But does this make any real sense?

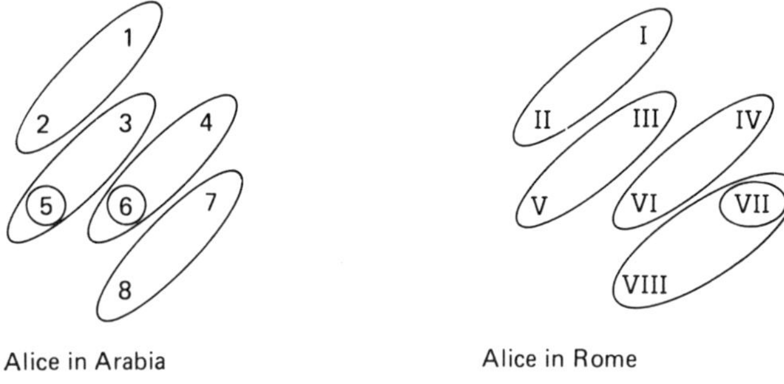

Figure 4.12

Consider the following alternative line of reasoning. Before anything happens, Alice does not "know" the informational structure of the world in which she lives. To be precise, she does not know whether or not the clergyman tells when there are exactly two dirty-faced ladies present. She may find this out. For example, she will learn that the clergyman *does* tell if state 5 occurs. But, if state 3 occurs, then it will remain possible for her that the clergyman does not tell. That is, she will have two *possible worlds* to take into account: the Arabic world on the left of figure 4.12 and the Roman world on the right. We therefore ought to be looking at figure 4.11 and calculating prob($\{3, \text{III}\}|\{3, 5, \text{III}\}$). This will be $2/3$ if Alice attaches equal prior probabilities to both the Arabic and the Roman worlds.

The point is that "ignoring ignorance" is not properly consistent with organizing uncertainties probabilistically. If Alice attaches a probability of $1/2$ in state 5 to the event E that her face is dirty given that Bertha is dirty and Cora is clean, then she is behaving as though she knew that the Roman world were impossible. But this conclusion is not one which anyone would wish to claim was a necessary consequence of her informational status. If she is not to explore the possibility that the world might be Roman, the most that one can reasonably assert about her conditional probability for the event E is that it lies between 1/2 and 1. That is, "ignoring ignorance" means that probabilistic judgments may be incomplete.

Holt (1988) comments on the interdependence of assumptions regarding probability orderings, informational structures and characteristics of knowledge and provides some references to the philosophical literature on the subject.

4.9 Common knowledge of the game

It is clear why it should be assumed that players *know* the structure of the game they are playing. But why should it be required that the game structure is *common* knowledge?

To make this point, a version of an example of Rubinstein (1988) will be described. The game confronts the players with a coordination problem similar to those considered by Lewis (1969) when introducing the notion of common knowledge. In such games, the players do not have a conflict of interests. Their problem consists in the fact that their freedom to communicate is restricted and it is therefore not necessarily easy for them to act as a "team" in coordinating on a joint plan of action.

The payoff matrix for the game to be considered is given in figure 4.13. At the beginning, the players do not know how the letters A and B are used to label the strategies. This is decided by tossing a weighted coin which comes down heads with probability 2/3 and tails with probability 1/3. If it were common knowledge that a head had appeared, then both players would, of course, choose strategy A. If it were common knowledge that a tail had

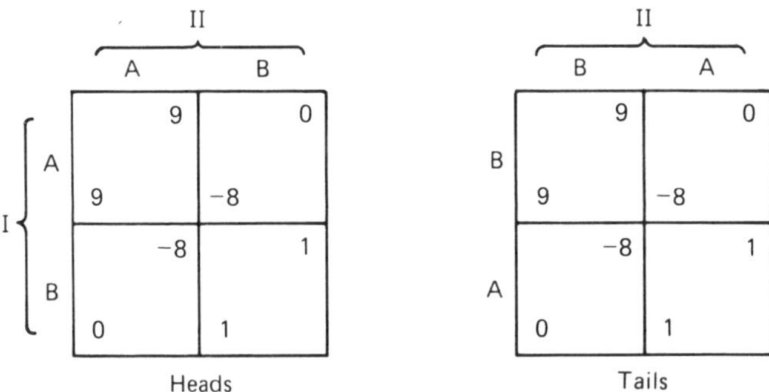

Figure 4.13

appeared, both would choose strategy B. However, only player I observes the fall of the coin.

Player I would like to inform player II of his information but can only communicate through the eletronic mail system as follows. If he sees a head he sends no message at all. If he sees a tail, he sends a message: Those familiar with the electronic mail system will know that messages fail to reach their destinations with some small probability $\varepsilon > 0$. If player II gets the message, she automatically sends an acknowledgement. If player I gets the acknowledgment, he automatically acknowledges the acknowledgment; and so on.

To study this informational set-up, the state space $\Omega = \{0, 1, 2, 3, \ldots,\}$ is introduced. A state ω in this space represents the number of messages sent. The event H that heads occurs is $H = \{0\}$. The event that tails occurs is $T = \{1, 2, 3, \ldots,\}$. Observe that

$$\omega \in (\text{everybody knows})^{\omega - 1} T \qquad (\omega = 1, 2, \ldots,)$$

For example, if $\omega = 3$, then tails occurred and player I knows it because he gets to see the coin. Moreover, player II knows and knows that I knows because she got a message saying so. Further, player I knows that II knows because he got an acknowledgment.

The question is whether players I and II can ever by sufficiently well informed to make it possible for them to choose strategy B when tails occurs. The answer is that they are *never* able to do so, no matter how large ω may be, if it is give that player I will always do the natural thing in those cases when heads occurs and choose A. If tails occurs, it may be that everybody knows that tails has occurred and everybody knows that everybody knows for 117 iterations. Nevertheless, they will still be unable to coordinate on strategy B.

To demonstrate this, some more apparatus is required. The players' information sets are shown in figure 4.14. For example, player II encloses 0 and 1 in the information set i_1, because either state is consistent with her receiving no message. It is necessary to say what each player would do at each of their information sets. It has already been noted that player I is to

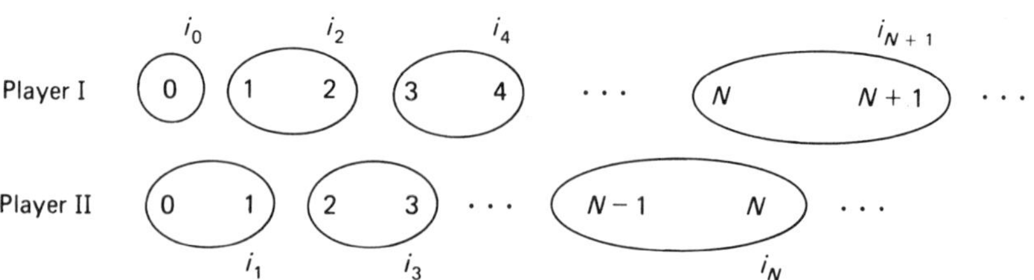

Figure 4.14

choose strategy A at i_0. To say what will happen elsewhere, probabilities need to be calculated.

Let $p(\omega)$ be the probability of state ω. Then $p(0)=2/3$, $p(1)=\varepsilon/3$ and $p(\omega+1)=(1-\varepsilon)p(\omega)(\omega=1,2,3,\ldots,)$. If $i=\{\omega,\omega+1\}$, then the conditional probabilities for ω and $\omega+1$, given that information set i has been reached, satisfy $p(\omega+1|i)/p(\omega|i)=p(\omega+1)/p(\omega)$. Thus

$$p(1|i_1)/p(0|i_1)=\varepsilon/2$$

$$p(n+1|i_{n+1})/p(n|i_{n+1})=1-\varepsilon \qquad (n=1,\ 2,\ 3,\ \ldots,)$$

Since player I chooses A at i_0, the least that player II can get from choosing A at i_1 is $x=9p(0|i_1)+0p(1|i_1)$. The most that can be got from choosing B at i_1 is $y=0p(0|i_1)+9p(1|i_1)$. Since $y/x=\varepsilon/2<1$, player II selects A at i_1.

The rest of the argument proceeds by induction. The induction hypothesis is that strategy A is chosen at information set i_n for a given $n\geqslant 1$. It is to be shown that A is also chosen at i_{n+1}. The least that can be got by choosing A is $x=1p(n|i_{n+1})+0p(n+1|i_{n+1})$. The most that can be got by choosing B is $y=-8p(n|i_{n+i})+9p(n+1|i_{n+1})$. Since $y/x=1-\varepsilon<1$, A is selected at i_{n+1}.

It is easy to see that $M(\omega)=\Omega$ in every state ω, and hence the outcome of the coin toss never becomes common knowledge and therefore neither does the structure of the game. This is the explanation proposed to account for the players' failure to coordinate successfully.

However, more than one loose end has been left. The first of these is taken up to some extent in the coming sections. Why should it be supposed that there will be successful coordination even if the game structure *were* common knowledge? For player I to know that a strategy is optimal, he needs to know something about what player II is going to do. To know what player II is going to do, he certainly needs to know something about what player II knows – which includes what player II knows about what player I knows, and so on. But he needs to know *more*. He needs to know how player II *uses* her knowledge. Game theorists typically seek to plug this gap with the assertion that it is "common knowledge that the players are rational." However, what this statement means in precise terms remains a vexed issue.

The second loose end concerns the question of *approximating* common knowledge. When tails occurs, the probability that of the event (everybody knows)$^N T$ in Rubinstein's example is nearly unity for each value of N when ε is sufficiently small. Does this not make T "nearly" common knowledge? If so, why is the players' behavior so far from what it would be if T were fully common knowledge?

One may seem to defuse the second question, as in Monderer and Samet (1988) and Stinchcombe (1988), by asking that only behavior which is *approximately* optimal be expected of the players. Monderer and Samet distinguish between *ex ante* ε-equilibria and *ex post* ε-equilibria.[6] With the former, one can require that A be chosen at i_0 and i_1 and B elsewhere. This calls for suboptimal play only at i_2. But i_2 occurs only with the small probability of $\varepsilon(1-\varepsilon)/3$.

If it were legitimate to use such *ex ante* ε-equilibria, the common knowledge difficulties with which this section is concerned would largely disappear. But, although i_2 may be reached with only very small probability, there seems no good reason to suppose that player I will choose very badly at i_2 on those few occasions when it *is* reached. *Ex post* ε equilibria require that each player make a response to the opponent which is within ε of the optimal available payoff at *every* information set.

When dealing with *ex post* ε-equilibria, it is necessary to look seriously at what constitutes "approximate common knowledge." Both Monderer and Samet (1988) and Stinchcombe (1988) tackle this issue by supplementing the knowledge operator K with a p-belief operator B_p defined so that B_pE may be interpreted as the set of states in which E is believed with probability at least p. One can then define common p-belief in much the same way as common knowledge and use the former as an approximation to the latter when p is close to unity. If there is approximate common knowledge in this sense, then Monderer and Samet show that there are *ex post* ε-equilibria close to the equilibria that prevail when there is full common knowledge.

4.10 Rationalizability

As noted in the preceding section, even when the structure of a game has been assumed to be common knowledge, there remains a yawning gap between the results of game theory and the tenets of Bayesian decision theory on which it is supposedly based. This gap has been traditionally plugged with *ad hoc* additions to Bayesian rationality concerning the use of equilibria in multi-person situations. These *ad hoc* additions are then defended with vague assertions about there being "common knowledge of rationality."

Kadane and Larkey (1982) are particularly outspoken in their criticism of this lacuna in the foundations of the subject. They argue that a Bayesian decision-maker maximizes utility given his subjective probability distribution and, if the resulting actions are not in equilibrium, so what? This view depends on the naive notion that Bayesian rationality somehow confers upon its adherents the capacity to "pluck their beliefs from the air." But what one believes must depend on what one knows, and, in game theory, the players know things about the other players.

Only recently have formal attempts been made to model the knowledge that a player has about the other players in a game *explicitly* (Aumann, 1987;

Bernheim, 1984, 1985; Brandenburger and Dekel, 1987b; Pearce, 1984; Reny, 1985; Tan and Werlang, 1988). These attempts all take account of the fact that a player's beliefs about the strategic choices of the other players in a game should not be entirely arbitrary.

Bernheim and Pearce's concept of "rationalizability" is the most conservative of the attempts. The only restriction imposed on beliefs is that everybody knows that everybody maximizes utility given their subjective probability distributions; and everybody knows that everybody knows; and so on. That is, it is common knowledge that the players are Bayesian rational. But the actual subjective probability distributions which the players hold are not assumed to be common knowledge.[7] In spite of the weakness of the assumptions, rationalizability can sometimes lead to a clear-cut prediction about the outcome of the game.

Consider the game illustrated in figure 4.15. Suppose that player I attaches a subjective probability of p to the event that player I will choose "top" and $1-p$ to the event that he will choose "bottom." Whatever the value of p, "right" cannot be a maximizing choice for player I because "left" is strictly better when $p > 1/3$ and "middle" is strictly better when $p < 3/4$. Since player I knows that player II is an optimizer, he therefore must assign probability 0 to the event that player II chooses "right." But then, whatever probabilities player I attaches to the other alternatives for player II, it is optimal for player I to choose "top." But player II knows that player I knows that player II is an optimizer. Thus player II knows that player I will necessarily choose "top." Hence player II chooses "left."

In two-player games, Bernheim and Pearce show that "rationalizable" strategies are those left after strictly dominated strategies are deleted from the game; and then strictly dominated strategies are deleted from the game that results; and so on. (The strategy "right" for player II in figure 4.15 is strictly dominated by an equal mixture of "left" and "middle.") This process of successively deleting dominated strategics gocs all the way back to Luce and Raiffa (1957). However, it is well known that it is only in rather special

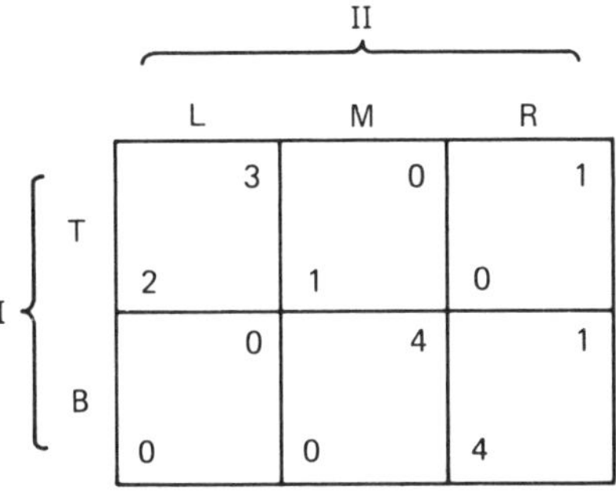

Figure 4.15

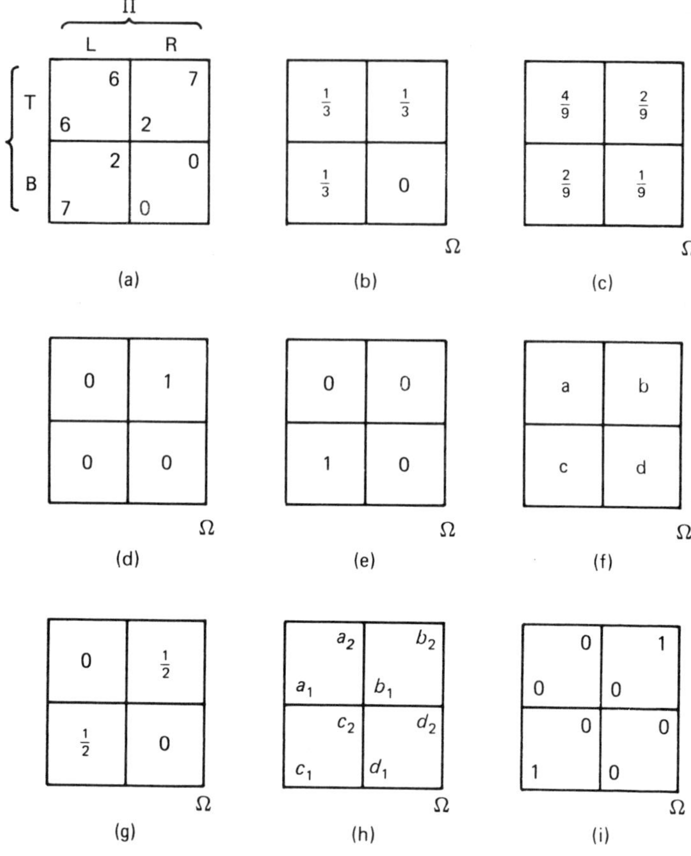

Figure 4.16

circumstances that the process generates a unique prediction. For the game of "chicken," illustrated in figure 4.16(a), the process has no bite at all. Every strategy is "rationalizable."

Bernheim (1984) regards this as a serious blow for the traditional equilibrium ideas of game theory. However, our feeling is that more is implicitly assumed by traditional game theory than Bernheim is willing to grant. One should therefore not be too surprised if, having thrown out the baby, one is left only with the bathwater.

4.11 Correlated equilibrium

In contrast with Bernheim, Aumann (1987) is willing to make quite strong common knowledge assumptions about players' beliefs in order to defend the notion of a correlated equilibrium.

Aumann's favorite example of a correlated equilibrium involves the game of "chicken" as illustrated in figure 4.16(a). Suppose that a random device selects one of the cells in figure 4.16(a) according to the probabilities indicated in figure 4.16(b). These probabilities are common knowledge, but player I is told only the row of the cell actually selected while player II is told only the column. Then player I and player II will have different but *correlated* information. Suppose that player I now uses the row reported to him as his strategy in "chicken" and player II does the same with the column reported to her. Then both players will be optimizing given the behavior of the other. For example, if player I is told "top," then he expects $6/2+2/2=4$ from playing "top" but only $7/2+0/2=3.5$ from playing "bottom." if he is told "bottom." then he expects 7 from playing "bottom" but only 6 from playing "top." The idea is, of course, very similar to that popularized by Cass and Shell (1983) under the name of "sunspot equilibrium." Aumann argues that "Bayesian rationality in games," if properly interpreted, is nothing other than the play of correlated equilibrium strategies. His argument makes this almost into a tautology. But to explain his point requires a little apparatus.

The specification of a two-player game in normal form consists of the players' strategy sets S_1 and S_2 and their Von Neumann and Morgenstern utility functions $u_1: S_1 \times S_2 \to \mathbb{R}$ and $u_2: S_1 \times S_2 \to \mathbb{R}$. Thus $u_1(s_1, s_2)$ is the utility that player I gets if he chooses strategy $s_1 \in S_1$ and player II chooses $s_2 \in S_2$. In the game of "chicken," $S_1 = \{\text{top, bottom}\}$, $S_2 = \{\text{left, right}\}$ and, for example, $u_2(\text{top, right}) = 7$.

What does Bayesian rationality require in such a situation? Suppose that, when $\omega \in \Omega$ occurs, players I and II observe $P_1(\omega)$ and $P_2(\omega)$ respectively, and that Bayesian rationality leads them to select[8] strategies $b_1(\omega) \in S_1$ and $b_2(\omega) \in S_2$. If each player's behavior is to be optimal in each state ω given the players' beliefs, then it must be the case that

$$\begin{aligned} \mathscr{E}u_1[b_1, b_2 | P_1(\omega)] &\geqslant \mathscr{E}u_1[d_1, b_2 | P_1(\omega)] \\ \mathscr{E}u_2[b_1, b_2 | P_2(\omega)] &\geqslant \mathscr{E}u_2[b_1, d_2 | P_2(\omega)] \end{aligned} \tag{4.2}$$

where d_1 and d_2 represent any alternative strategy to that prescribed by Bayesian rationality.[9]

The essential point to be gleaned from this discussion is simply that requirement (4.2) *is* an equilibrium condition. Aumann's (1987) gloss is to observe that (4.2) implies the condition for a correlated equilibrium provided that, in each state $\omega \in \Omega$, the players *know* the Bayesian rational strategy. For example, in the game of "chicken," let TOP denote the set of those $\omega \in \Omega$ for which Bayesian rationality requires player I to choose "top." Then to know his Bayesian rational strategy, player I will need to know when the event TOP occurs and, similarly, when the event BOTTOM occurs. As far as "chicken" is concerned, he can discard any other information he might have and just retain the information partition {TOP, BOTTOM} of Ω. Similarly, player II need only retain the partition {LEFT, RIGHT}. But then we have the picture of figure 4.16(b) and hence Bayesian rationality has generated for us the correlated equilibrium discussed earlier, provided that the players' prior beliefs (i.e. the beliefs they have *before* observing anything) satisfy

$$\text{prob}(\text{TOP} \cap \text{LEFT}) = \text{prob}(\text{TOP} \cap \text{RIGHT}) = \text{prob}(\text{BOTTOM} \cap \text{LEFT}) = 1/3$$

This is the first explicit mention of the players' prior beliefs on Ω. What do the players know about these? There is no difficulty in supposing that players know their *own* prior beliefs on Ω. They are therefore in a position to check that the behavior required by Bayesian rationality does indeed maximize their expected utility, provided the other players do not deviate. But to check that the others will not deviate, a player needs to know something about the prior beliefs of the players. Otherwise, it would not be possible to check that the behavior required by Bayesian rationality maximizes *their* utility also. Aumann (1987) argues that we actually have no choice but to assume that all the prior subjective probability distributions are, in fact, *common knowledge.* He reasons that, if player j were to learn that $\omega \in \Omega$ had occurred but continued to entertain two possibilities for the subjective probability attached by player i to the state ω, then ω would not be an all-inclusive description of the state of the world. An *all-inclusive* description would include a description of the probability that player j attaches to that description. Thus, everybody must know everybody's *prior* beliefs and, by a similar argument, everybody must know that everybody knows, and so on. We share the unease that the reader will probably feel about the self-reference built into this argument, for the reasons already discussed in section 4.6. For us, the assumption of common knowledge of prior probability distributions therefore remains something which needs defense.

We have seen how common knowledge assumptions provide the glue which holds together Aumann's defense of correlated equilibrium in Bayesian context. Some taxonomy is now appropriate. This depends on the answers to the following questions.

1 Are the players *independent*?
2 Do the players have *common* priors?

To say that the players are independent means that all the players believe that their own information is statistically independent of that received by the other players. Thus, for example, in "chicken," both players' priors will satisfy $\text{prob}(\text{TOP} \cap \text{LEFT}) = \text{prob}(\text{TOP}) \times \text{prob}(\text{LEFT})$. Think of the description of $\omega \in \Omega$ as including the fall of a weighted coin observed only by player I and the fall of an independent weighted coin observed only by player II.

To say that the players have common priors is to say that their prior beliefs on Ω are the same. This is *not*, of course, implied by the hypothesis that their priors are common knowledge. The players may "agree to disagree" about what the prior beliefs should be.

We begin with the case when it is simultaneously true that the players choose independently and that they have common priors. These are the traditional assumptions of noncooperative game theory and the conclusion is the traditional conclusion: namely, that the result will be a *Nash equilibrium.* For example, if the players' prior beliefs are as indicated in figure 4.16(c), then the result is a Nash equilibrium for "chicken." An observer will see player I choose "top" with probability 2/3 and "bottom" with probability 1/3, while player II independently chooses "left" with probability 2/3 and "right" with probability 1/3. Each of these "mixed strategies" is an optimal reply to the other. Given player II's behavior, player I will get $6 \times 2/3 + 2 \times 1/3 = 14/3$ from the choice of "top" and $7 \times 2/3 + 0 \times 1/3 = 14/3$ from choice of "bottom." He is therefore indifferent between these two alternatives and therefore content to use any mixture of them. Similarly for player II.

"Chicken" has two other Nash equilibria. These involves only "pure strategies." In the first, player I always chooses "top" and player II always chooses "right." (The players' common prior is then as in figure 4.16(d).) For the second Nash equilibrium in pure strategies, player I always chooses "bottom" and player II always chooses "left." (The players' common prior is then as in figure 4.16(e).)

Note that for Nash equilibrium, in contrast with the general case of a correlated equilibrium, the players' posterior beliefs about each other's *choice of strategy* are common knowledge. If we had not begun with a discussion of correlated equilibrium, this might have been a more natural characterization of Nash equilibrium. (See Brandenburger and Dekel (1987b) and Aumann (1988) for a different view.) In the correlated equilibrium supported by the prior of figure 4.16(b), for example, player I's posterior belief about player II's choice of strategy depends on whether I observes TOP or BOTTOM, but II does not know which of these I observes.

Before leaving the subject of Nash equilibrium, there is a point to be made about the relevance of *mixed* equilibria. Economists sometimes reject these out of hand on the grounds that economic agents simply do not randomize when making decisions. But such a view depends on adopting a rather naive interpretation of what a mixed Nash equilibrium is. One advantage of working strictly in terms of an underlying state space Ω is that a more sophisticated interpretation lies immediately on the surface. In the mixed Nash equilibrium for "chicken" described above, neither player actively randomizes. Both simply see themselves as choosing deterministically, given their information. The random element arises from uncertainty *in the mind of the other player* about what that choice will be. It may well be that real-life economic agents do not consciously randomize when making decisions, but they will nearly always be conscious of being uncertain about what other agents will decide. No a priori case therefore exists for the wholesale rejection of mixed equilibria in the economic context.

The case when the players' choices may be correlated, but prior beliefs are common, is that of correlated equilibria. In "chicken," the correlated equilibrium described earlier is only one of many. In general, the prior subjective probability distributions which support a correlated equilibrium are described by a simple system of linear inequalities. Referring to figure 4.16(f), it is easy to check that the appropriate inequalities for "chicken" are

$$\begin{aligned} 6a + 2b &\geqslant 7a \\ 7c &\geqslant 6c + 2d \\ 6a + 2c &\geqslant 7a \\ 7b &\geqslant 6b + 2d \end{aligned}$$

Nash equilibria are special cases of correlated equilibria and it is not difficult to see that any convex combination of Nash equilibrium outcomes is achievable as a correlated equilibrium. (Just let the players jointly observe a random device which selects a Nash equilibrium and then require them to take whatever independent actions are necessary to implement that Nash equilibrium.) For example, the common prior to figure 4.16(g) gives the two pure Nash equilibrium outcomes for "chicken," each with probability $1/2$. But note that the resulting payoff of $7/2 + 2/2 = 4.5$ for each player is not so good as the expected payoff of $6/3 + 2/3 + 7/3 = 5$ that each player gets with the correlated equilibrium supported by the common prior of figure 4.16(b). The expected outcome with the latter correlated equilibrium is *not* obtainable as a convex combination of Nash equilibrium outcomes. For a prettier example, see Moulin and Vial (1978).

For the case when priors are not common, consider figure 4.16(h). The numbers a_1, b_1, c_1, and d_1 are subjective probabilities for player I and the numbers a_2, b_2, c_2, and d_2 are subjective probabilities for player II. All these numbers are common knowledge but the players "agree to disagree." Figure 4.16(i) is a special case chosen to highlight the fact that something counter-intuitive is perhaps involved in such situations. With priors as in figure 4.16(i), both players believe it is certain that they will come away from the game with their maximum payoff of 7.

It is pleasant to be able to round off this section by pointing out that Bernheim and Pearce's rationalizability is not so distant from Aumann's correlated equilibrium after all. Brandenburger and Dekel (1978b) have shown that rationalizability can be recast in an equilibrium mold. In the two-player case, a player can rationalize a payoff if and only if it is a component of the vector of payoffs from what Aumann calls a "subjective correlated equilibrium." By this is meant an equilibrium of the type we have just been discussing: namely, one in which the players may agree to disagree and use commonly known but different priors. This result emphasizes the importance of the assumption of common priors in determining which solution concept is to be regarded as the "correct" expression of Bayesian rationality.

Finally, what of independence when the priors are not common? On this point, we want only to mention that, in discussing "rationalizability," we have restricted attention always to the case $n = 2$, because Bernheim and Pearce make independence assumptions which become relevant when $n \geqslant 3$ but which it would be pedantic to describe here.

4.12 Updating beliefs

Section 4.4 describes how the acquisition of new information may lead decision-makers to update their information partitions. Such refining of their information partitions will be accompanied by an updating of their prior beliefs. The study of this phenomenon is clearly very important if one hopes to get some sort of handle on markets in which speculation is an important element.

The current discussion begins with Aumann's "agreeing to disagree" result (which is proposition 4.6). The agents are endowed with common knowledge of a common prior. Each then receives some private information. Their posterior probabilities for an event F then become common knowledge. Aumann's result is that these posterior probabilities must necessarily be equal.

An example of Geanakoplos and Polemarchakis (1982) serves to show that this result leaves various stones unturned. In this example, Alice and Bertha have a common prior that attaches equal probabilities to each element of $\Omega = \{1, 2, 3, \ldots, 9\}$. Their information partitions are $\{A_1, A_2, A_3\}$ and $\{B_1, B_2, B_3\}$ respectively as illustrated in figure 4.17. For example, $P_A(4) = A_2$

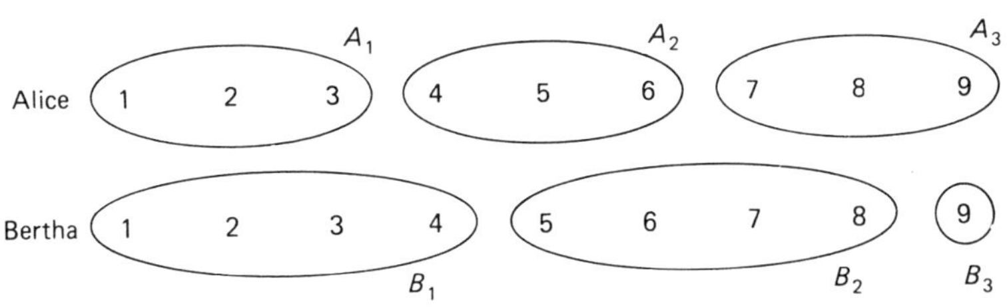

Figure 4.17

and $P_B(9) = B_3$. The meet M of P_A and P_B (section 4.4) satisfies $M(\omega) = \Omega$ for all $\omega \in \Omega$. It follows from proposition 4.5 that the only event that can be common knowledge when it occurs is Ω. Thus, if

$$E = \{\omega: \text{prob}[F | P_A(\omega)] = q_A\}$$

is common knowledge as required in the hypotheses of Aumann's result, then $E = \Omega$. This means that Alice *always* attaches probability q_A to the event F whatever her private information. Similarly Bertha always attaches probability q_B to F. Their private information is therefore irrelevant to the event F and hence Aumann's result is empty for this example.

Consider, however, the case when $F = \{3, 4\}$ and the following sequence of truthful interchanges between Alice and Bertha takes place. Suppose that $\omega = 2$. Then, Alice observes $P_A(2) = A_1 = \{1, 2, 3\}$. She then communicates her posterior probability of 1/3 that the event F has taken place to Bertha. Bertha observes $P_B(2) = B_1 = \{1, 2, 3, 4\}$ and so, before receiving Alice's message, has a posterior probability of 1/2 for F. On getting Alice's message, she deduces that Alice has observed A_1 or A_2 but cannot deduce anything else. Since she knew this already, she leaves her posterior probability for F alone. The next step is for her to report this to Alice. Alice deduces that Bertha must have observed B_1. Since she knew this already she leaves her posterior for F alone and reports this to Bertha. But *this* tells Bertha a great deal. She knows Alice observed A_1 or A_2. If Alice had observed $A_2 = \{4, 5, 6\}$, then the information that Bertha observed $B_1 = \{1, 2, 3, 4\}$ would have conveyed the fact to Alice that $\omega = 4$ and so Alice would have reported a posterior probability of 1 for F. Since she did not, Bertha deduces that Alice observed $A_1 = \{1, 2, 3\}$ and hence now announces a posterior probability of 1/3 for F. After this, all future announcements of the posterior probability of F are 1/3. It has become common knowledge that the event $A_1 = \{1, 2, 3\}$ has occurred and it is common knowledge that $\text{prob}(F | A_1) = 1/3$.

Alice and Bertha therefore reach a *consensus* in respect of their beliefs about F. That is, the probabilities they attach to F finally become common knowledge and hence are equal (by proposition 4.6). But more is true in this particular case. They also end up with the *same* information, namely that A_1 has occurred. This need not always happen. Geanakoplos and Polemarchakis provide a second example in which the agents' posterior beliefs about an event are common knowledge, but *not* their full information about the event. The commonly known consensus beliefs are then *not* what they would be if the two agents pooled *all* their information.

The example is illustrated in figure 4.18. The state space is $\Omega = \{1, 2, 3, 4\}$ and the common prior is that all states are equally likely. When $\omega = 1$, the consensus probability for the event $F = \{1, 4\}$ is 1/2, but pooling the total information would reveal that F is certain.

While there are situations in which it makes practical sense to imagine

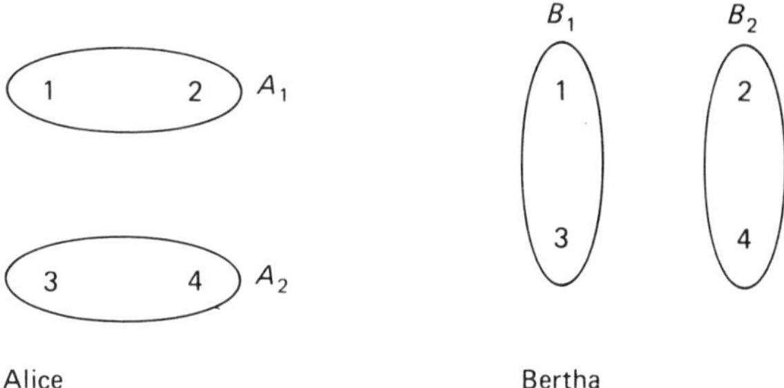

Figure 4.18

agents directly and truthfully reporting their posterior probabilities for some event in the manner examined above, a more likely scenario is that only some statistic of individual beliefs becomes a public event. An example is when asymmetrically informed agents come to a market to trade. The trading process causes private information to be aggregated into a public statistic, such as a price or quantity. If agents recompute their beliefs on the basis of the value of this public statistic, and then further prices or quantities are announced, will the agents eventually reach a situation of common knowledge and hence consensus in their beliefs? This question has an obvious relevance to rational expectations equilibria but arises in numerous other contexts also.

There is a considerable literature in statistics on the reconciliation of differing expert opinions (e.g. Dalkey, 1969). The so-called "Delphi technique" is a commonly discussed example. The experts are envisaged as offering predictions of the likelihood of some event based on their private information. The average of their predictions is announced publicly. The experts then use this information to revise their predictions. This leads to the announcement of a new average and so on. Pari-mutuel betting has similar characteristics.

McKelvey and Page (1986) have demonstrated how common knowledge can arise through the publication of such aggregate statistics. (A number of people have noticed that simple proofs of the result are possible, e.g. Brandenburger and Geanakoplos (1986).) We shall look only at the "Delphic" special case of their result. Each of n agents computes his or her posterior probability $q_i(\omega)$ for an event F. Thus, originally, $q_i(\omega) = \text{prob}[F \mid p_i(\omega)]$. Then the statistic

$$\phi(\omega) = \frac{1}{n} \sum_{i=1}^{n} q_i(\omega)$$

is published. Each agent revises his or her posterior $q_i(\omega)$, using the value of the published statistic. These posteriors are then averaged and the new statistic is published; and so on. The result is that, after a finite number of iterations of this process, the agents' posteriors must become common knowledge and so consensus is achieved.

An example (taken from McKelvey and Page, 1986) may be of assistance. Suppose that Alice, Bertha, and Cora are experts who face a state space $\Omega = \{1, 2, 3, 4, 5\}$. The common prior is that each state is equally likely. The initial private information is provided by the information partitions illustrated in figure 4.19. The experts are interested in the probability of the event $F = \{1, 2, 3\}$.

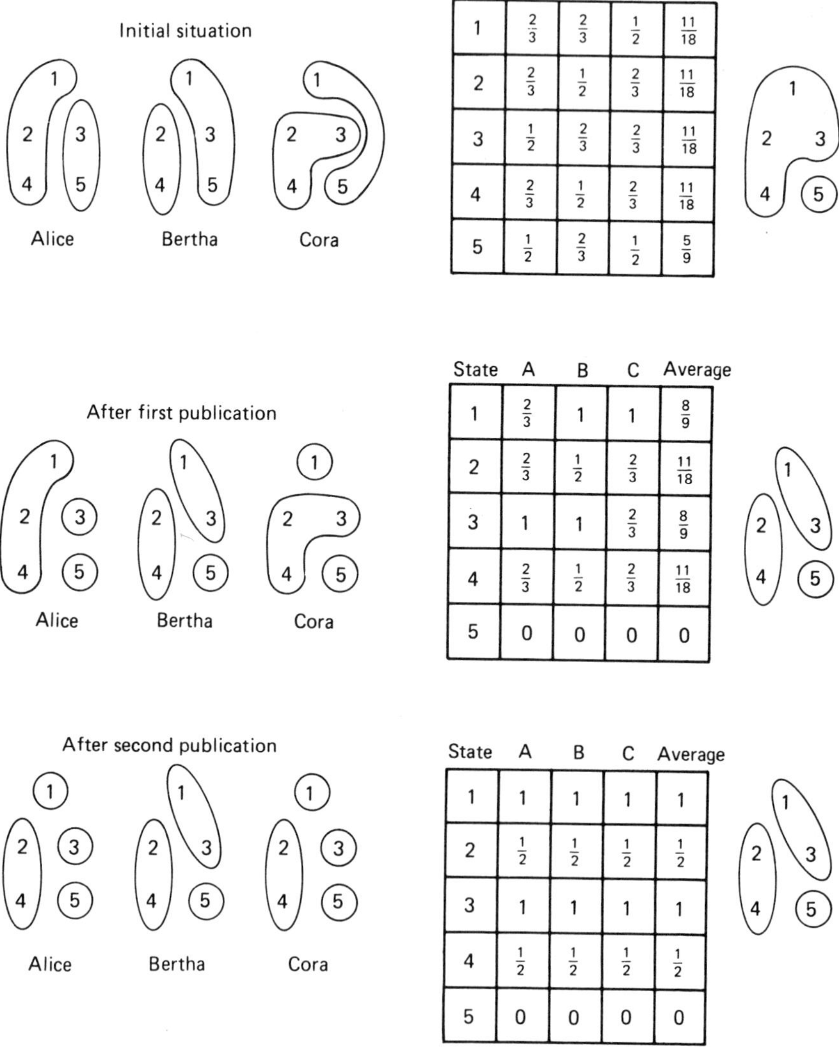

State	A	B	C	Average
1	$\frac{2}{3}$	$\frac{2}{3}$	$\frac{1}{2}$	$\frac{11}{18}$
2	$\frac{2}{3}$	$\frac{1}{2}$	$\frac{2}{3}$	$\frac{11}{18}$
3	$\frac{1}{2}$	$\frac{2}{3}$	$\frac{2}{3}$	$\frac{11}{18}$
4	$\frac{2}{3}$	$\frac{1}{2}$	$\frac{2}{3}$	$\frac{11}{18}$
5	$\frac{1}{2}$	$\frac{2}{3}$	$\frac{1}{2}$	$\frac{5}{9}$

State	A	B	C	Average
1	$\frac{2}{3}$	1	1	$\frac{8}{9}$
2	$\frac{2}{3}$	$\frac{1}{2}$	$\frac{2}{3}$	$\frac{11}{18}$
3	1	1	$\frac{2}{3}$	$\frac{8}{9}$
4	$\frac{2}{3}$	$\frac{1}{2}$	$\frac{2}{3}$	$\frac{11}{18}$
5	0	0	0	0

State	A	B	C	Average
1	1	1	1	1
2	$\frac{1}{2}$	$\frac{1}{2}$	$\frac{1}{2}$	$\frac{1}{2}$
3	1	1	1	1
4	$\frac{1}{2}$	$\frac{1}{2}$	$\frac{1}{2}$	$\frac{1}{2}$
5	0	0	0	0

Figure 4.19

Suppose that the true state is $\omega = 1$ so that Alice, Bertha, and Cora calculate their posterior probabilities for F to be $2/3$, $2/3$, and $1/2$ respectively. The average of these assessments, $(2/3 + 2/3 + 1/2)/3 = 11/18$, is then published. This gives Alice no new information but informs Bertha and Cora that $\{5\}$ did not occur. Hence the experts' revised posteriors are $2/3$, 1, and 1. Thus a statistic of $(2/3 + 1 + 1)/3 = 8/9$. At this point, Alice can conclude that $\{2, 4\}$ did not occur since this would have resulted in a second-round announcement of $(2/3 + 2/3 + 2/3)/3 = 11/18$. She therefore revises her posterior probability of F to 1 and consensus is achieved.

This example deserves a more careful discussion. The table on the first line of figure 4.19 shows the probabilities that the three experts will assign to the event F in each possible state of the world. The information partition on the right of this table is that generated by the publication of the average of these probabilities. That is, it is the information partition that would be held by an individual who is informed of the averages published so far, but of nothing else. The second and third lines of figure 4.19 show how these information partitions are refined as updated averages of the probabilities assigned by the experts to F are published. Observe that consensus is achieved in all states of the world after only two publications.

This expanded version of the argument illustrates a point that is true in general: namely, that the value of the current average must be common knowledge after consensus is reached (Brandenburger and Geanakoplos, 1986). An expert updates her information partition by taking the coarsest common refinement of her current information partition and the partition generated by the latest publication. After consensus has been achieved, this updating will produce no change. It follows that the partition generated by the latest publication must then be at least as coarse as the meet of the experts' partitions. But then the fact that the average is equal to any particular number is a common truism. The value of the average is therefore common knowledge.

Such simple models are reminiscent of Keynes's "beauty contest" story. Keynes drew attention to newspaper competitions of the time in which the aim was not to identify the most beautiful young lady whose photograph was supplied but to predict the way the voting would go on this matter. The aim, in both cases, is to draw attention to certain aspects of the problem of speculation in financial assets. What models like that of McKelvey and Page (1986) add to the Keynesian picture is a formal identification of the basic difficulty. The trading activities of rational agents with common knowledge of a common prior, who all receive different pieces of private information, must be expected to lead to an eventual consensus in beliefs, after which no rational basis for speculation will exist. One might rationalize continued speculation by supposing that the agents are in continuous receipt of new items of private information about an environment assumed to be constantly changing (although we do not know of work in which this is attempted within a formal common knowledge framework). But is this the way things really are? Or is it simply that speculators are actually "agreeing to disagree"? (Varian (1987) further discusses these issues.)

4.13 Are common knowledge assumptions realistic?

In brief, our view is that it is precisely in those applications which lean most heavily on common knowledge assumptions, whether this is explicitly recognized or not, that the assumptions are least realistic. We have particularly in mind those models in which a great deal can happen as time passes and a number of agents, each of significant size, seek to learn what has happened and to adapt their behavior to the circumstances. One might summarize what we have to say with the observation that it seems to us painfully naive to suppose that Bayesian updating captures more than a tiny part of what is involved in genuine learning. Whatever the answer to the problem of scientific induction may be, it will surely involve more than the trivial algebraic manipulation called Bayes's rule.

Savage's (1954) theory, in which he synthesized Von Neumann and Morgenstern's expected utility theory with the subjective probability ideas of Ramsey, De Finetti and others, is entirely and exclusively a *descriptive* theory of consistent behavior. It has nothing to say about how decision-makers acquired the beliefs ascribed to them: it asserts only that, if the decisions taken are consistent, then the decision-makers act *as though* they were maximizers of expected utility relative to a subjective probability distribution. (Objections to the definition of "consistent" in the theory are possible but this is not our point. In particular, the objection that "real people" are sometimes inconsistent is irrelevant to a theory of rational behavior. People often get their sums wrong. But nobody argues that we should therefore change the rules of arithmetic.)

The "Bayesian blunder" is to suppose that Savage's passive descriptive theory can be reinterpreted as an active prescriptive theory at negligible cost. Obviously, a sensible decision-maker will be unhappy about inconsistencies. A naive Bayesian therefore assumes that it is enough to assign prior beliefs to a decision-maker and then forget the problem of where beliefs come from. This was the attitude we adopted in the preceding sections in sketching the nature of the orthodox approach to common knowledge. Consistency then forces any new information that may transpire to be incorporated into the system by Bayesian updating – i.e. a posterior belief is deduced from the prior belief using Bayes's rule. The naiveté does *not* consist in using Bayes's rule, whose validity as a piece of algebra is not in question. It lies in supposing that the problem of where the priors come from can be quietly shelved.

Some authors even explicitly assert that rationality somehow *endows* decision-makers with priors and hence the problem does not exist at all.

Savage (1954) had a considerably more complex view. He did argue that his descriptive theory could be of practical assistance in helping decision-makers form their beliefs. His point was that rational agents would not rest if they found inconsistencies in their belief systems. Luce and Raiffa (1957, p. 302) expound Savage's view as follows:

> Once confronted with such inconsistencies, one should, so the argument goes, modify one's initial decisions [about beliefs] so as to be consistent. Let us assume that this jockeying – making snap judgments, checking on their consistency, modifying them, again checking on consistency, etc. – leads ultimately to a bona fide, *a priori* distribution.

For what follows, we need to expand on this quotation. What is at issue is *why* a rational decision-maker should want to be consistent. After all, scientists are not consistent, on the grounds that it is not clever to be consistently wrong. When surprised by data that shows current theories to be in error, they seek new theories which are inconsistent with the old theories. This is what *genuine* learning is like. Consistency, from this point of view, is only a virtue if the possibility of being surprised can be eliminated somehow. How does Savage see this situation being achieved?

A person who makes judgments in a reasonable way will presumably prefer to make judgments when he or she has more information rather than less. A decision-maker might therefore begin to tackle the problem of constructing an appropriate belief system by asking: for every conceivable possible course of future events, what would my beliefs be after experiencing them? Such an approach automatically discounts the impact that new knowledge will have on the basic model being used to determine beliefs – i.e. it eliminates the possibility that the decision-maker will feel the need to alter this basic model after being surprised by a chain of events whose implications had not previously been considered. Next comes the question: is this system a *contingent* beliefs consistent? If not, then the decision-maker may examine the relative *confidence* that he or she has in the "snap judgments" he or she has made and then adjust the corresponding beliefs until they *are* consistent. With Savage's definition of consistency, this is equivalent to asserting that the adjusted system of contingent beliefs can be deduced, using Bayes's rule, from a single prior. It is therefore true, in this story, that the final "massaged" posteriors can be deduced formally from the final "massaged" prior using Bayes's rule. This is guaranteed by the use of a complex adjustment process which operates until consistency is achieved. As far as the massaged beliefs are concerned, Bayes's rule therefore has the status of a *tautology*, like $2+2=4$. Together with the massaged prior, it serves essentially as an indexing system which keeps track of the library of massaged posteriors. It would perhaps be wrong to argue that there is no learning going on when an index system is used to locate a book in a library, but it will be clear that, if the above story is to be believed, then the real learning takes place *during the massaging process.* Notice that what happens during the real learning process is that a massaged *prior* is deduced from a set of primitive *posteriors.* To pursue the analogy given above, the actual learning should be thought of as assembling the right books to go in the library, the *final* step being the provision of a suitable index.

Sophisticated Bayesians, among whom we would like to count ourselves, sometimes follow Harsanyi in signaling that some analog of Luce and Raiffa's jockeying procedure is required by asserting that the Bayesian theory applies only to "closed universe" or small-world problems, i.e. to problems in which all potential surprises can be discounted in advance (section 4.6). Suitable examples are to be found in the small decision trees with which looks on Bayesian decision theory for students of business administration are illustrated. Naive Bayesians make no such qualifications. For them, something which ia not formalized in Bourbaki-type mathematics does not exist and hence no account need be taken of it all.

The implications of taking a sophisticated view of Bayesian theory are serious enough when only one decision-maker is involved. In sections 4.6 and 4.7, and again in section 4.11, we considered the orthodox practice of asserting that the states of the world $\omega \in \Omega$ are to be *all-inclusive.* This allows various tricks of the trade to be practised on the unwary. However, whatever else may be clear, a problem in which *all* future histories, without exception, have to be taken into account is not one for which the massaging process described earlier makes any practical sense whatsoever. If Bayesian conclusions are not to be abandoned altogether, it is therefore necessary to be very much more circumspect about *how* inclusive a state of the world is taken to be.

However, although all this is related to what comes next, it is not the main point we wish to make. The main point concerns what Aumann (1976) calls the "Harsanyi doctrine": namely, that priors should be taken to be common, to which Aumann adds the rider that it should be common knowledge that the priors are common. The orthodox defense need not be spelled out in detail since it is closely related to the idea of the "original position" as popularized by Rawls. Individuals are thought of entering the world with a mind which is a *tabula rasa.* Since rational individuals in this "original position" will all have the same information, how could they adopt different priors?

Obviously, such a defense of the common prior assumption will not survive the interpretation of Savage's theory offered above. However, one can recast the defense as follows. Imagine all the players constructing their beliefs, before there is any action, using the Luce and Raiffa jockeying process. How could beliefs come to be commonly held under such circumstances? Each player would need to be able to argue with confidence that he would make the *same* subjective judgments as any other player under *all* circumstances provided that he or she were to contemplate precisely the same contingencies under precisely the *same* conditions. (Harsanyi (1977) refers to a related notion as "extended sympathy.") In this case, each player could regard the other players, in their aspect of information processors, simply as surrogates of himself. The massaging process would therefore generate a common prior.

But this new defense asks us to swallow a great deal more than the old defense. Both require us to think of individuals beginning with an identical kit of information-processing tools and no information at all. Differences between individuals them emerge entirely as a result of different information that they receive as time passes. If the kit of tools is conceived of as containing only Bayes's rule and nothing else, then the idea that we can mimic the reaction of other players to their information does not seem fraught with difficulty. But once the massaging process is taken into account, we are faced with a tool-kit of unknown composition. Not only that, people learn about how to learn as they gain experience. A person's tool-kit after the passage of some time will not therefore be the same as his or her original tool-kit. Even if one is prepared to grant the doubtful proposition that we "know" what tools we have in our current kit (i.e. we know the mechanism by means of which we currently make subjective judgments), it does not follow that we can distinguish those tools that we have acquired through experience from those with which we were equipped originally. And probably it does not even make sense to suppose that we can itemize those tools we would have had in our current kit if our experience of the world had been other than it was. But, without this sequence of increasingly unlikely requirements, the new defense is unworkable. This is not to say that the defense may not have value as a useful approximation in simple enough situations, but only that it seems unwise to elevate its conclusions into holy writ.

We continue this last point with some remarks on extensive-form games. In sections 4.10 and 4.11, we looked only at normal-form games (in which all the action can be thought of as taking place at a single instant). But, in extensive-form games, time enters the picture. During the play of such a game, the players may receive evidence that the hypotheses they need to sustain the Harsanyi doctrine, or some other doctrine, are *false*. Figure 4.20 is the game-tree of an example adapted by Reny (1985) from a similar example of Rosenthal (1981) (see also Basu, 1985). Reny calls the game "take it or leave it." A philanthropist arrives with $\$10^n$ unsure of which of two institutions to endow. Alice and Bertha represent the two institutions. The philanthropist makes them play the followng game which may proceed through n stages. If the kth state is reached, the philanthropist will have placed $\$10^k$ on the table. If k is odd, Alice can take the money and so end

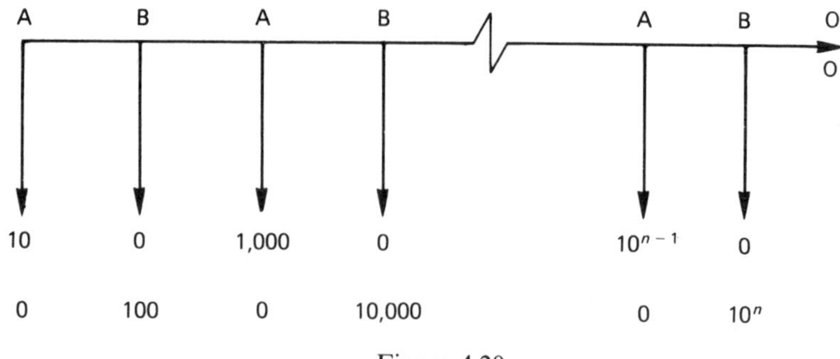

Figure 4.20

the game. If k is even, Bertha can take the money and end the game. If the money is not taken and $k < n$, then the philanthropist multiplies the amount available by ten and the game proceeds to the kth stage. If $k = n$, the philanthropist leaves in disgust taking the money with him.

Reny's (1985) point is very simple. He notes that "rationalizability" (and most other game theory "solution concepts") requires Alice to take the money at step 1. But what would Bertha then deduce aboute Alice if Alice did *not* take the money? The natural answer is that Alice is not a rationalizer. Whan then is she? The "rationalizing" theory offers no clue. But suppose Bertha decides that, whatever sort of person Alice is, if she left $10 on the table at step 1, then it is very likely that she will leave $1,000 on the table at step 3. In this case player II will leave $100 at step 2, expecting to be able to claim $10,000 at step 4. However, if Alice anticipates all this, it will not be optimal for her to "rationalize" at step 1 because she will be passing up the chance of being able to claim $1,000 at step 3.

Such discussions are absent from traditional game theory. In Selten's (1975) concept of perfect equilibrium, any deviation by players from the "Bayesian rational" course of action is attributed to small random errors. The players' hands are said to "tremble" slightly in selecting their actions, so that there is always a small chance that they will do the wrong thing by "mistake," although they always intend to do the right thing. In chess, for example, we are supposed to explain a long sequence of bad moves by asserting that the opponent always meant to move the correct piece but continually picked the wrong one up by mistake. More often, however, a formalism is adopted within which it is difficult, or impossible, even to express the relevant issues so that they then do not "need" to be considered at all. One advantage of being explicit about what is being assumed about the players, as discussed in section 4.10, is that such evasions become very difficult to sustain.[10]

Progress on these issues must presumably await advances in the theory of bounded rationality (although Aumann, 1988, has a proposal for cutting the

Gordian knot). Binmore (1987–8 – this volume, chs 5 and 6) reviews some of the issues.

4.16 Conclusion

What does all this mean for the study of common knowledge? Do the criticisms and reservations of the preceding section mean that its continued investigation is fruitless? This would not be a good conclusion to draw. Firstly, our criticisms are directed at applications of the theory which hopelessly overload its currently rather tenuous underpinnings. In small-scale applications, however, such as normal-form games with a small number of strategies, these criticisms are not necessarily relevant. Even in the much more doubtful large-scale applications, there is comfort to be drawn from the fact that, without an explicit understanding of the nature of common knowledge, it would not even be possible to properly appreciate where the inadequacies lie in the theories which are current.

NOTES

1 There are various difficulties with the raw version of Littlewood's story. One might respond to A's question by asking why B is still laughing given that she can analyze the situation just as easily as A.

2 The first criterion implies the second because, if $T=(ek)^{\infty}T$, then

$$(ek)T=(ek)(ek)^{\infty}T=(ek)^{\infty}T=T$$

The second criterion implies the first because, if $T=(ek)T$, then $(ek)T\subseteq(ek)^2T$ by (K1). But $(ek)^2T\subseteq(ek)T$ by (K2). Thus $T=(ek)T=(ek)^2T$. Similarly, $T=(ek)^nT$ for all n, and so $T=(ek)^{\infty}T$.

3 In symbols

$$P(\omega)=\bigcap_{\omega\in T}T=\bigcap_{\omega\in KE}E$$

where T ranges over all truisms and E over all events.

4 The union of two truisms, S and T, is a truism because $S\cup T=KS\cup KT\subseteq K(S\cup T)\cup K(S\cup T)=K(S\cup T)$.

5 Formally, this means that the event

$$E=\{\omega\colon \text{prob}[F|P_{\text{A}}(\omega)]=q_{\text{A}}\}$$

is common knowledge. Thus $M(\omega_0)\subseteq E$ by proposition 4.5. Suppose $Q_j=P_{\text{A}}(\omega)$. Then $\omega\in P_{\text{A}}(\omega)\subseteq M(\omega_0)\subseteq E$. Hence $\text{prob}(F|Q_j)=\text{prob}[F|P_{\text{A}}(\omega)]=q_{\text{A}}$.

6 Stinchcombe works with *ex ante* ε-equilibria.

7 A natural criticism is that Savage's (1954) theory constructs Von Neumann–Morgenstern utilities and subjective probabilities *simultaneously*. At first sight it therefore seems odd that one should be the subject of common knowledge

but not the other. Brandenburger and Dekel (1987b) present an expanded model with private information in which this objection does not bite. See also the discussion at the end of section 4.11.

8 A significant evasion is concealed in this argument. The assumption that Bayesian rationality leads to a specific recommendation for behavior rather than merely a set of constraints on behavior is left undefended.

9 Shin (1988) suggests that, with a truly all-inclusive description of a state, (4.2) is not appropriate since part of what one observes is the action one actually takes. Hence the $P_i(\omega)$ on the left of the inequalities should not be the same as that on the right.

10 Fudenberg, Kreps, and Levine (1988) and Dekel and Fudenberg (1987) address the issues raised in this paragraph in a more precise context.

REFERENCES

Aumann, R. 1976: "Agreeing to disagree," *Annals of Statistics* 4, 1236–9.

Aumann, R. 1987: "Correlated equilibrium as an expression of Bayesian rationality," *Econometrica* 55, 1–18.

Aumann, R. 1988: "Irrationality in game theory (preliminary notes)", unpublished, Hebrew University of Jerusalem.

Bacharach, M. 1985: "Some extensions to a claim of Aumann in an axiomatic model of knowledge," *Journal of Economic Theory* 37, 167–90.

Bacharach, M. 1987: "When do we have information partitions?" unpublished, Christchurch College, Oxford.

Basu, K. 1985: "Strategic irrationality in extensive games," unpublished, Institute of Advanced Study, Princeton, N.J.

Bernheim, D. 1984: "Rationalizable strategic behavior," *Econometrica* 52, 1007–28.

Bernheim, D. 1985: "Axiomatic characterizations of rational choice in strategic environments," unpublished, Department of Economics, Stanford University.

Binmore, K. 1984: "Equilibria in extensive games," *Economic Journal* 95, 51–9.

Binmore, K. 1987–8: "Modelling rational players I and II," *Economics and Philosophy* 3 and 4, 179–214 and 9–55.

Binmore, K., and Brandenburger, A. 1988: "Common knowledge and game theory," ST/ICERD Discussion Paper 88/167, London School of Economics.

Brandenburger, A. 1986: "The role of common knowledge assumptions in game theory," to appear in *The Economics of Information, Games, and Missing Markets*, ed. F. Hahn.

Brandenburger, A., and Dekel, E. 1985: "Hierarchies of beliefs and common knowledge," Research Paper 841, Graduate School of Business, Stanford University, Calif.

Brandenburger, A., and Dekel, E. 1987a: "Common knowledge with probability 1," *Journal of Mathematical Economics* 16, 237–45.

Brandenburger, A., and Dekel, E. 1987b: "Rationalizability and correlated equilibria," *Econometrica* 55, 1391–402.

Brandenburger, A., and Geanakoplos, J. 1986: "Common knowledge of summary statistics," unpublished, Cowles Foundation, Yale University, Conn.

Cass, D., and Shell, K. 1983: "Do sunspots matter?" *Journal of Political Economy* 91, 193–227.

Cave, J. 1983: "Learning to agree," *Economic Letters* 12, 147–52.

Dalkey, N. 1969: "The Delphi method: an experimental study of group opinion," Discussion Paper, Rand Corporation, Santa Monica, Calif.

Dekel, E., and Fudenberg, D. 1987: "Rational behavior with payoff uncertainty," unpublished, Department of Economics, University of California at Berkeley, Calif.

Fudenberg, D., Kreps, D., and Levine, D. 1988: "On the robustness of equilibrium refinements," *Journal of Economic Theory* 44, 354–80.

Gädenfors, P. 1975: "Qualitative probability as an intentional logic," *Journal of Philosophical Logic* 4, 171–85.

Geanakoplos, J. 1988: "Common knowledge, Bayesian learning and market speculation with bounded rationality," unpublished, Yale University, Conn.

Geanakoplos, J., and Polemarchakis, H. 1982: "We can't disagree forever," *Journal of Economic Theory* 28, 192–200.

Gilboa, I. 1986: "Information and meta-information," Working Paper 3086, Foerder Institute for Economic Research, Tel Aviv University.

Halpern, J. (ed.) 1986: *Theoretical Aspects of Reasoning about Knowledge*. Los Altos, Calif.: Morgan Kaufmann.

Harsanyi, J. 1967–8: "Games with incomplete information played by 'Bayesian' players," Parts I–III, *Management Science* 14, 159–82.

Harsanyi, J. 1977: *Rational Behavior and Bargaining Equilibrium in Games and Social Situations*. Cambridge: Cambridge University Press.

Holt, D. 1988: "Models of knowledge and information," unpublished, University of Michigan, Mich.

Kadane, J., and Larkey, P. 1982: "Subjective probability and the theory of games," *Management Science* 28, 113–20.

Kaneko, M. 1987: "Structural common knowledge and factual common knowledge," RUEE Working Paper 87–27.

Lewis, D. 1969: *Conventions: A Philosophical Study*. Cambridge, Mass.: Harvard University Press.

Littlewood, J. E. 1953: *Mathematical Miscellany*, ed. B. Bollobas. London: Cambridge University Press, 1986.

Luce, R., and Raiffa, H. 1957: *Games and Decisions*. New York: Wiley.

McKelvey, R., and Page, T. 1986: "Common knowledge, consensus, and aggregate information," *Econometrica* 54, 109–27.

Mertens, J. F., and Zamir, S. 1985: "Formulation of Bayesian analysis for games with incomplete information," *International Journal of Game Theory* 14, 1–29.

Milgrom, P. 1981: "An axiomatic characterization of common knowledge," *Econometrica* 49, 219–22.

Milgrom, P., and Stokey, N. 1982: "Information, trade, and common knowledge," *Journal of Economic Theory* 26, 17–27.

Monderer, D., and Samet, D. 1988: "Approximating common knowledge with common beliefs," Center for Mathematical Studies in Economics and Decision Science, Northwestern University.

Moulin, H., and Vial, J. P. 1978: "Strategically zero-sum games: the class of games whose completely mixed equilibria cannot be improved upon," *International*

Journal of Game Theory 7, 201–21.
Nielsen, L. 1984: "Common knowledge, communication, and convergence of beliefs," *Mathematical Social Sciences* 8, 1–14.
Pearce, D. 1984: "Rationalizable strategic behavior and the problem of perfection," *Econometrica* 52, 1029–50.
Reny, P. 1985: "Rationality, common knowledge, and the theory of games," unpublished, Department of Economics, Princeton University.
Rosenthal, R. 1981: "Games of perfect information, predatory pricing, and chain-store paradox," *Journal of Economic Theory* 25, 92–100.
Rubinstein, A. 1988: "A game with 'almost common knowledge': an example," forthcoming in *American Economic Review*.
Rubinstein, A., and Wolinsky, A. 1988: "A comment on the logic of 'agreeing to disagree' results," unpublished, Department of Economics, Northwestern University.
Samet, D. 1987: "Ignoring ignorance and agreeing to disagree," Discussion Paper 749, J. L. Kellogg. Graduate school of Management, Northwestern University.
Savage, L. 1954: *The Foundations of Statistics*. New York: Wiley.
Schelling, T. 1960: *The Strategy of Conflict*. Cambridge, Mass.: Harvard University Press.
Selten, R. 1975: "Reexamination of the perfectness concept for equilibrium points in extensive games," *International Journal of Game Theory* 4, 25–55.
Shin, H. 1986: "The structure of common knowledge," unpublished, Nuffield College, Oxford.
Shin, H. 1987: "Logical structure of common knowledge, I and II," unpublished, Nuffield College, Oxford.
Shin, H. 1988: "A comment on Aumann's definition of Bayes-rationality," unpublished, Nuffield College, Oxford.
Stinchcombe, M. 1988: "Approximate common knowledge," unpublished, University of California at San Diego, Calif.
Tan, T., and Werlang, S. 1985: "On Aumann's notion of common knowledge: an alternative approach," Working Paper 85–26, University of Chicago, Ill.
Tan, T., and Werlang, S. 1988: "The Bayesian foundations of solution concepts of games," *Journal of Economic Theory* 45, 370–91.
Varian, H. 1987: "Differences of opinion in financial markets," unpublished, Department of Economics, University of Michigan, Mich.

5 Modeling rational players: part I

For the Schooles find ... no actuall Motion at all; but because some Motion they must acknowledge, they call it Metaphoricall Motion; which is but an absurd speech.

Hobbes, *Leviathan*

5.1 Introduction

Game theory has proved a useful tool in the study of simple economic models. However, numerous foundational issues remain unresolved. The situation is particularly confusing in respect of the noncooperative analysis of games with some dynamic structure in which the choice of one move or another during the play of the game may convey valuable information to the other players.[1] Without pausing for breath, it is easy to name[2] at least ten rival equilibrium notions for which a serious case can be made that here is the "right" solution concept for such games.

It is not the purpose of this chapter to survey the various contenders (although section 5.2 does contain some expository material for those who do not follow the game theory literature). The purpose is rather to question some of the shibboleths that underlie current game-theoretic research and that seem to me to obstruct further progress.

This is a much revised two-part version of a previous working paper, "Modeling Rational Players" (ST/ICERD discussion paper 86/133). Some of the material of the original paper has been discarded and the remainder has been expanded, revised, and refined. However, it remains a philosophical piece about the foundations of game theory. The essential point is that the traditional ideal or axiomatic approach needs to be abandoned in favor of a constructive or algorithmic approach.

In retrospect, the most striking feature of the paper is the manner in which the final version has turned out to be a mirror of the preoccupations of Harsanyi and Selten (1988) in their recent book on the "tracing procedure." I continue to be dubious about the details of the solutions they propose to the problems they raise, but I am no longer dubious about the necessity of making a frontal attack on these problems, nor about the proposition that any solution will necessarily be complicated and arbitrary to some extent. If the paper serves to convince others also, it will perhaps have served a useful purpose.

Both this paper and the working paper were written with the support of NSF grant number SES-8605025.

At root, the difficulties are philosophical in that they arise from the manner in which the nature of the problems to be resolved is perceived. Usually the problems are framed in abstract mathematical terms and then attacked à la Bourbaki. Such a definition–axiom–theorem–proof format closes the mind, the aim being to exclude irrelevancies so that attention can be focused on matters of genuine importance. My contention is that the conventional approach misses this aim, not only by leaving unformalized factors which matter, but also by introducing formal requirements that cannot be defended operationally except in terms of mathematical elegance or simplicity.[3] What we have to say on this subject is closely related to the distinction Simon (1976) draws between *substantive rationality* and *procedural rationality*. He comments that economists confine their attention almost exclusively to the former (and psychologists to the latter).

Discussions of substantive rationality take place in an essentially *static* framework. Thus, equilibrium is discussed without explicit reference to any dynamic process by means of which the equilibrium is achieved.[4] Similarly, prior beliefs are taken as given, without reference to how a rational agent acquires these beliefs. Indeed, all questions of the procedure by means of which rational behavior is achieved are swept aside by a methodology that treats this procedure as completed and reifies the supposed limiting entities by categorizing them axiomatically.[5] In particular, it is usually taken for granted that the notion of a *perfectly rational agent* can be assigned a sharp and unambiguous meaning. In fact, game theoretic analyses often implicitly assume that it is meaningful to assert considerably more: namely that "it is common knowledge that all the players are perfectly rational."[6] This implicit axiom confronts the analyst with the necessity of dealing with counterfactuals of the type: suppose a perfectly rational player carried out the following irrational sequence of acts .... Such considerations are unavoidable. A perfectly rational player will not deviate from his equilibrium strategy. But a profile of strategies is in equilibrium because of what *would* happen if a player *were* to deviate. It is the various attempts to deal with such counterfactuals that motivate the quotation from Hobbes at the head of this chapter.

A digression on the meaning of the word *rational* is necessary at this point. A glance at any dictionary will confirm that economists, firmly entrenched in the static viewpoint described above, have hijacked[7] this word and used it to mean something for which the word *consistent* would be more appropriate. Such an inversion is certainly useful as a rhetorical device. Who can argue with someone's advocating the rational course of action? But, in so far as scientific inquiry is concerned, such ploys can only be a source of unnecessary confusion. Consider, for example, the title of Sen's (1976) paper "Rational fools." It could be argued that a new word should be sought for what rational used to mean (and still does mean to the layman) before it was hijacked. But there seem no suitable alternatives. Myerson (1984) proposes *intelligent*, but the world has seen no shortage of intelligent madmen. In any case, a *rational decision process* will be understood in this chapter to refer to the *entire* reasoning activity that intervenes between the receipt of a decision stimulus and the ultimate decision, *including* the manner in which the decision-maker forms the beliefs on which the decision is based. In particular, to be rational will *not* be taken to exclude the possible use of the scientific method.

Such an approach forces rational behavior to be thought of as essentially *algorithmic*. This makes it natural to seek to model a rational player as a suitably programmed computing machine. Such a viewpoint is, of course, orthodox among mathematicians in respect of deductive mathematical reasoning. But there is a specter at this feast. What of Gödel's incompleteness theorem? Loosely expressed, this tells us that *all* reasoning systems[8] are *necessarily* imperfect in the sense that any such system can be fitted inside a bigger and better system (see, for example, Kline, 1980). Before dismissing such considerations as absurdly esoteric, it should be remembered that Gödel's proof depends on a self-reference argument and that self-reference is intrinsic in traditional game-theoretic analyses because of their dependence on chains of reasoning that begin "If I think that you think that I think ...."

It has been argued elsewhere (e.g. Binmore, 1984) that such considerations cast doubt on the extent to which "perfect rationality" is a genuinely meaningful notion. Further arguments to this effect are also offered later in this chapter. None of these arguments can be said to be conclusive,[9] but they do serve to indicate the possibility that traditional analyses may have reified a nonexistent object. If this possibility is admitted, it becomes necessary to assign rationality a role similar to that of an *uncompleted infinity* as envisaged in classical (pre-Cantorian) mathematical analysis, i.e. a useful metaconcept specifically *not* treated as a formal entity in the system under study. Any formal model for a rational player will then necessarily be imperfect in the sense that, whatever model is chosen, there will be situations for which the model does not perform adequately and hence is capable of being improved. There are, of course, numerous technical problems involved in working with *bounded rationality*. It is also uncomfortable to entertain the prospect of employing models whose structure is necessarily arbitrary to some extent. But there are some conceptual rewards to be considered. To quote Selten (1975, p. 35):

> There cannot be any mistakes if the players are absolutely rational. Nevertheless, a satisfactory interpretation of equilibrium points in extensive games seems to require that the possibility of mistakes is not completely excluded. This can be achieved by a point of view which looks at complete rationality as a limiting case of incomplete rationality.

The mistakes are required to escape the counterfactuals referred to earlier. Thus, if a rational player chooses an irrational act, it may be attributed to some sort of blunder. But the manner in which the making of mistakes is modeled matters a great deal. Selten (1975) attributes mistakes to uncorrelated random errors that intervene between the decision to take a certain action and the action itself (thc so-called tremblind-hand explanation). The approach we are currently advocating would instead attribute the primary source of mistakes to imperfections *within the reasoning process itself*, wherever this is possible.

This brings us to a second and rather different criticism of the traditional approach to game theory. Not only are abstractions introduced that do not necessarily admit an operational referent; at the same time, operationally relevant factors are abstracted away altogether. This, in itself, is not necessarily indivious. However, the Bourbaki ethos makes it inevitable that factors that are not taken account of formally are not taken account of at all. In particular, the traditional static approach to game theory results in equilibrium being discussed without any reference to the equilibrating process[10] by means of which the equilibrium is supposedly achieved. To my mind, this is like trying to discuss animal anatomy without a Linnaean classification scheme. Or, to take a more homely example, like trying to decide which of the roots of a quadratic equation is the "right" solution without reference to the context in which the quadratic equation has arisen. The point here is that the same formal game might receive *different* analyses depending on the environment from which it has been abstracted, i.e. that the analysis of a game may require more information than is classically built into the formal definition of a game.

A consequence of the failure to begin with a preliminary informal classification of the interesting environments within which games are played is that confusion reigns about what criteria are reasonable in selecting equilibria.[11] Criteria appropriate in one environment are uncomfortably yoked with criteria appropriate to quite different environments, while other criteria relevant in the first environment are neglected altogether.

A classification of game-playing environments, and specifically of types of equilibrating process, needs to take account of numerous issues, of which perhaps the most difficult concern the manner in which players process data (i.e. the modeling of their reasoning processes) and the manner in which data that serve as input to this process are generated and disseminated. If, as seems inevitable to me, out-of-equilibrium behavior is to be treated in terms of mistakes, then information on these issues is essential in order that a judgment be possible on which of the many possible types of mistake is the most likely in a given situation.

In chess, for example, it simply does not make sense, given the environment in which it is normally played, to attribute bad play by an opponent to a sequence of uncorrelated random errors in implementing the results of a flawless thinking process.[12] This is not to say that it is impossible that the opponent's elbow may have been repeatedly jogged while he or she was in the act of moving a piece: only that this is a comparatively unlikely explanation. Rather than resort to such a trembling-hand explanation, one would look for some *systematic* error in the way in which the opponent analyzes chess positions. The detection of such a systematic error will have important implications about the opponent's expected future play, whereas the observation of a trembling-hand error will have no such implications. The trembling-hand explanation therefore enjoys immense advantages in terms of mathematical simplicity and elegance. But this does not necessarily make it the "right" explanation for the environment in question. Nor does the argument offered above suggest that it may not be the "right" explanation in other environments. In fact, it seems particularly well suited to situations in which it is appropriate to model the players as very simple stimulus–response machines whose behavior has become tailored to their environment as a result of ill-adapted machines' having been weeded out by some form of evolutionary competition. Some scenarios from animal biology provide good examples[13] (Selten, 1983).

The word *eductive*[14] will be used to describe a dynamic process by means of which equilibrium is achieved through careful reasoning on the part of the players. Such reasoning will usually require an attempt to simulate the reasoning processes of the other players. Some measure of pre-play communication is therefore implied, although this need not be explicit. To reason along the lines "If I think that he thinks that I think ..." requires that information be available on how an opponent thinks.

The word *evolutive* will be used to describe a dynamic process by means of which equilibrium is achieved through evolutionary mechanisms. It is intended to include not only the very long-run processes studied by evolutionary biologists (Maynard Smith, 1982), but also medium-run processes in which the population dynamics are not necessarily based on genetic considerations (e.g. Friedman and Rosenthal, 1984), as well as the very short-run processes by means of which markets achieve clearing prices (e.g. Marshak and Selten, 1978; Moulin, 1981). The linking consideration is that adjustment takes place as a result of *iterated* play by *myopic* players.

Of course, the distinction between an eductive and an evolutive process is quantitative rather than qualitative. In the former, players are envisaged as potentially very complex machines (with very low operating costs) whereas, in the latter, their internal complexity is low.[15] It is not denied that the middle ground between these extremes is more interesting than either extreme. However, only isolated forays have so far been made into this area (e.g. Neymann, 1985; Rubinstein, 1985; Abreu and Rubinstein, 1986).

This chapter is concerned with rational players, and it is therefore eductive processes that are chiefly relevant. But a strictly eductive environment is seldom encountered in the real world.[16] What intuitions we may glean about "reasonable" properties of equilibria from empirical observation are therefore contaminated by the intrusion of evolutive factors. My view is that much of the confusion concerning equilibrium ideas in game theory can be traced to this fact. Intuitions which belong in an evolutive theory have been defended from an eductive position. In offering unorthodox views on eductive issues, I am therefore not necessarily denying the validity of such intuitions: I am merely suggesting that they be properly recognized as evolutive.

This long preamble may now make it possible to come clean without provoking immediate hostility. Leaving aside the idiosyncratic tracing procedure of Harsanyi and Selten, it is perhaps Kohlberg and Mertens (1983, 1986) who have made the most systematic attempt to build a sound theory of rational equilibrium selection. As basic tenets they offer the principles of "backwards induction," "the successive elimination of dominated strategies," and "equivalence of games which can be mapped into one another by 'strategically inessential transformations.'" In what follows, it is heretically argued that *none* of these is acceptable in a strictly eductive context, nor is *any* approach based only on trembles in the *external* environment[17] (although this is not to exclude such trembles as epiphenomena). Since the equilibrating process is internal to the players, it is *inside* the reasoning process of the players than an explanation for deviancy should properly be sought.

It is appreciated that the acceptance of such views more or less guarantees that no neat and tidy eductive theory of games is feasible. But game theorists need not feel apologetic. A neat and tidy theory would imply an incidental resolution of the problem of scientific induction in a particularly aggravated form.[18] And this is obviously too much to hope for, given our current state of knowledge.

The organization of the current chapter is as follows. Section 5.2 reviews some basic ideas for the noncognoscenti. Nothing at all complex is required to appreciate the points to be made later. Sections 5.3 and 5.4 are devoted to examples which are intended to challenge various orthodoxies about the eductive analysis of games. Section 5.5 is an attack on the notion of "perfect rationality" itself. Finally, section 5.6 seeks to explain why some of these issues are not even seen as issues by those raised in the orthodox Bayesian school.

The current part of the study is therefore almost entirely critical in content. The second part of the study addresses itself to more positive questions. Given the criticisms of part I, what are the issues that have to be resolved before a game-playing rational agent can be modeled? In brief, it is argued that a computing machine capable of some measure of introspection is required and that such machines will necessarily be arbitrary to some degree.

5.2 Some game theory

In this section, an example of Kohlberg (see, for example, Kohlberg and Mertens, 1983, or Kreps and Wilson, 1982a) is used to illustrate some basic ideas from noncooperative game theory. The example will be referred to as Kohlberg's dalek example because of the shape of its game tree (figure 5.1a).

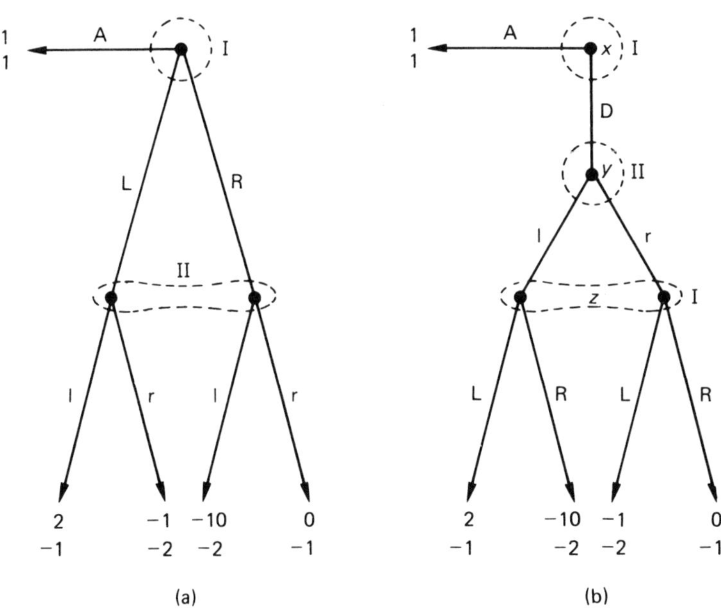

Figure 5.1

Traditionally, game theories are classified as cooperative or noncooperative. In cooperative theories, the players are assumed to be able to communicate freely before the play of the game and to be able to conclude *binding* agreements on what strategy each is to use. Noncooperative theories vary in the extent to which pre-play communication is assumed possible, but always insist that any pre-play agreements are definitely *not* binding.[19] Nash (1951) observed that games with *no* explicit pre-play communication (which we call contests) should be regarded as fundamental since, in principle, the various maneuvers possible during any pre-play communications can be modeled as formal moves in a formal "negotiation game" that it would then be appropriate to analyze as a contest. Of course, some background pool of information must be commonly held[20] if any worthwhile analysis is to be possible. An essential, therefore, is some *implicit* communication that may be attributed to such factors as a shared cultural heritage. Even if Nash's proposal is rejected as impractical, it remains the case that the analysis of contests is of interest. In any case, this paper and its successor confine their attention to such games.

The formal rules of a game can be specified in detail with the aid of game tree. Figure 5.1(a) is the game tree for Kohlberg's dalek example. Classical theory holds that the game tree of figure 5.1(b) is strategically equivalent and, for expository purposes, it is preferable to work with this instead.

The game illustrated in figure 5.1(b) has two players labeled I and II. (We adopt the convention that odd players are male and even players are female.) The arrows point to the possible outcomes of the game. These are labeled with the utilities that the players attach to each outcome. Utilities are intended in the sense of Von Neumann and Morgenstern, which means that players act *as though* seeking to maximize *expected* utility. The first move is located at the node labeled x. Here player I chooses between across or down (A or D). If he chooses D, then it becomes player II's turn to choose at node y. She chooses between left or right (l or r). Whichever she chooses, it then becomes player I's turn to choose between left or right (L or R). He makes this choice *in ignorance of* the decision just made by player II. This is indicated by enclosing *two* nodes (that reached via the decision history (D,l) and that reached via (D,r)) within the *information set* labeled z. A player making a choice at a node within an information set is assumed to know only that one of the nodes within the information set has been reached but not which of these nodes it is. A *play* of the game is a sequence of decisions connecting the initial node with a final outcome. One possible play of the game of figure 5.1(b) is the sequence ⟨D,l,R⟩, which leads to an outcome to which player I assigns utility -10 and player II assigns -2. Another play consists simply of ⟨A⟩. This leads to an outcome to which both players assign utility 1. Chance moves can be incorporated into such a set-up where necessary by assigning nodes to a mythical player 0 who makes each choice with a predetermined probability.

Notice that a game must have an initial node.[21] For this reason, the game of figure 5.1(a) has *no* subgames. The game of figure 5.1(b) has precisely one proper subgame: that obtained by deleting node x. If node y is deleted also, the remaining configuration has no initial node and therefore does not qualify as a game.

A *pure strategy* for a player specifies the choice the player plans to make at *every* information set assigned to him or her by the rules of the game. The pure strategies for player I in figure 5.1(b) may be characterized[22] as (A,L), (A,R), (D,L), (D,R), and those for II simply as l and r. A player is said to use a *mixed strategy* if the decision about what pure strategy to use is delegated to a random device. Such a player determines only the probability with which each pure strategy is chosen. After Selten (1975), it has become customary to confine attention to games of *perfect recall* (in which no player ever forgets information he or she previously knew). In such games mixed strategies may be replaced, without loss of generality, by *behavioral strategies.* A behavioral strategy specifies, for each information set, the probability with which each action available to the player at that information set is to be selected. In essence, a behavioral strategy *decentralizes* the randomization required by a mixed strategy.

Von Neumann and Morgenstern believed that games with some dynamic structure could be reduced, without losing generality, to *static games.* The term static game indicates a game in which all action is telescoped into a single instant. At this single instant, the players simultaneously and irrevocably make all decisions that need to be made. In particular, Von Neumann and Morgenstern would have regarded the *extensive-form* game of figure 5.1(b) as equivalent to what they called its "normal form." The latter is represented twice in figures 5.2(a) and 5.2(b). Clearly the tabular form of figure 5.2(b) is the more convenient rendition.

A *Nash equilibrium* for a game is a strategy profile (one, possibly mixed, strategy for each player) that has the property that each player's strategy choice is an optimal response to the strategy choices of the other players. Nash (1951) proved that all finite games have at least one Nash equilibrium. For a *static* contest analyzed eductively it is usually held to be a *sine qua non* that a solution concept be at least a Nash equilibrium.[23,24] The Nash equilibria for the game of figure 5.1(b) are conveniently extracted from the normal form as given in figure 5.2(b). The Nash equilibria in pure strategies are [(A,L),r], [(A,R),r], and [(D,L),l]. It is also a Nash equilibrium if player I uses any probabilistic mixture of (A,L) and (A,R) and player II uses r. No other Nash equilibria exist.

Selten's (1975) view, that Von Neumann and Morgenstern were mistaken in identifying a game with its normal form, is now widely accepted.[25] In particular, Selten argued that all Nash equilibria for a game like that of figure 5.1(b) are not viable candidates for the solution of the game. The reason is that the assumption that players can proceed by making an irrevocable commitment to a strategy before the game begins is unwarranted. If they can make such a commitment, then this commitment possibility should be formally modeled as part of a larger game. If they cannot make such a commitment (a more realistic hypothesis for most situations), then they remain free to change their minds about what to do as the game proceeds. In the latter case, Selten argued that the reasons for restricting attention to Nash equilibria in the whole game also apply in any subgame, *whether or not* that subgame is actually reached. Nobody will be deterred from deviating from a prescribed equilibrium action if the deterrent depends on rational players' planning to play irrationally. A Nash equilibrium which induces

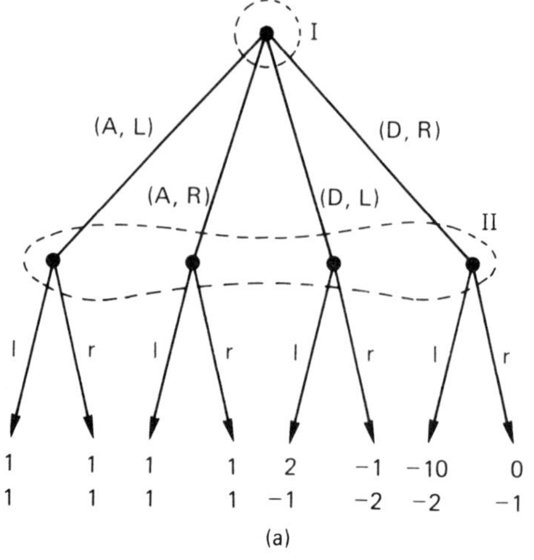

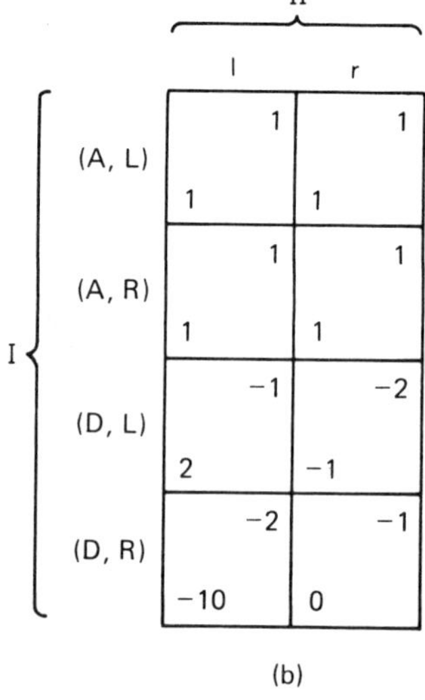

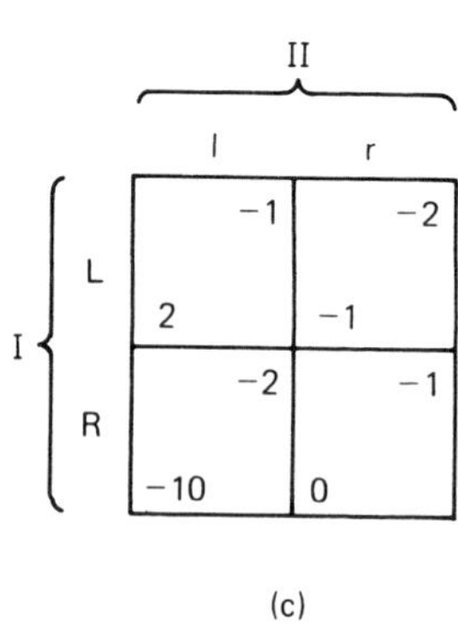

Figure 5.2

Nash equilibrium play in all subgames of the original game is said to be *subgame-perfect*.[26]

The game of figure 5.1(b) has only one subgame whose normal form is given in figure 5.2(c). Its Nash equilibria are (L,l), (R,r), and a mixed equilibrium in which player I uses L with probability 1/2 and player II uses l with probability 1/13. It follows that all the Nash equilibria for the original game can be rejected as not being subgame-perfect except for [(A,R),r] and

[(D,L),l]. Consider, for example, [(A,L),r]. This is rejected because if node *y* were to be reached then player II would not choose r if she were aware that player I was planning to choose L.

Selten's basic idea is that a player's contingent plans should make sense whether or not the contingencies are actually realized in the game. Clearly, this ought to be true, in some sense, at each information set rather than just in each subgame. (For an example with no subgames but for which such considerations are crucial, see the "horse" game of section 5.5.) Selten's (1975) *trembling-hand perfect* equilibrium generalizes subgame-perfect equilibrium with this aim in view. Each (behavioral) strategy in a trembling-hand perfect equilibrium is required to be an optimal response not only to the other strategies in the profile but to small perturbations of these strategies. The perturbations are obtained by supposing that players' hands tremble (independently at each information set) so that, although they may decide to press the button for action A, there is always a positive probability that the button actually pressed will be that for some other action B. This uncertainty can be built into the rules of the game by introducing new chance moves at appropriate places. A trembling-hand equilibrium can then be viewed as the limit of a sequence of Nash equilibria in such "perturbed games" as the perturbations are allowed to become negligible.[27]

Kreps and Wilson (1982a) offer a notion they call *sequential equilibrium.* This is only marginally weaker than a trembling-hand equilibrium, but is technically easier to work with and is couched in terms closely comparable with those of subgame-perfect equilibria. A sequential equilibrium is an "assessment" profile. This consists of a behavioral strategy profile as before together with a "belief" profile. The latter species a probability distribution for each information set. These are the probabilities that a player will assign to the nodes in an information set should he or she could be called upon for a decision at the information set. It is required that the beliefs be *consistent* and that strategic plans always be optimal given these beliefs and the strategic plans of the other players. In figure 5.1(b), for example, the strategy profile [(A,R),r] constitutes a sequential equilibrium provided that the associated belief profile attaches probability 1 to the right-hand node of information set *z* and 0 to the left-hand node. If these probabilities are reversed, a belief profile is obtained that supports [(D,L),l] as a sequential equilibrium.

In the second of these examples of a sequential equilibrium, player 1 may deduce his beliefs at *z* from his prediction that player II will choose l at node *y*. However, in the first example, player I's choice of A at *x* guarantees that *z* will *not* be reached. His beliefs at *z* are therefore predicated on a zero probability event. Much of the debate about the choice of equilibrium concepts has centered on precisely this issue: what is it reasonable or plausible to believe given the observation of a zero probabiltity event?

Kolmogorov (1950) answers this question for the general case of computing prob($A|B$) when prob(B) = 0 by recommending the use of the limiting value of prob($A|B_n$) for a suitable sequence $\langle B_n \rangle$ of events with prob(B_n) > 0 such that $B_n \to B$ as $n \to \infty$. When prob(B_n) > 0, Bayes's rule may be applied. This is the route followed implicitly by Selten's trembling-hand idea and explicitly by Kreps and Wilson in defining what they mean by consistent beliefs. But Kolmogorov warns, with instructive examples, that the "wrong" result will be obtained if the "wrong" sequence $\langle B_n \rangle$ is employed. However, the question of which is the "right" sequence $\langle B_n \rangle$ is not a mathematical question: it is a question of good modeling judgment. In section 5.1, it was suggested that the elbow-jogging basis for the construction of $\langle B_n \rangle$ used by Selten and by Kreps and Wilson does not represent a good judgment when a game like chess is to be analyzed *eductively*.

This last point is pursued in section 5.3. The current section continues by reviewing some further game-theoretic ideas. Section 5.1 mentions three Ur-principles for rational play proposed by Kohlberg and Mertens (1983, 1986). "Backwards induction" is incorporated in Selten's subgame-perfect equilibrium (or, more generally, in Kreps and Wilson's sequential notion). To compute subgame-perfect equilibria in a game, it is first necessary to know the subgame-perfect equilibria of all proper subgames. Backwards induction is simply the process of beginning with the final subgames (subgames with no proper subgame) and then successively calculating the equilibria of larger subgames until the original game is reached. The term "dynamic programming" is used in other contexts. Their second principle is the "successive deletion of dominated strategies." This principle is relevant to the question of plausible beliefs in the dalek example of figure 5.1(b).

In figure 5.2(b), pure strategy (A,L) for player I *strictly dominates* his pure strategy (D,R). The requirement is that the former give a *strictly* better payoff than the latter *whatever* strategy is used by player II. It is usually argued (but see section 5.3) that rational players will never use a strictly dominated strategy. On deleting the row labeled (D,R) in figure 5.2(b) a table is obtained in which player II's pure strategy 1 *weakly* dominates her pure strategy r. If the latter is deleted on the grounds that it does her no harm to do so given that player I will not choose (D,R), a simple 3 × 1 table remains. For this table, player I's pure strategy (D,L) strictly dominates his other remaining strategies. An argument has therefore been given for selecting the equilibrium [(D,L),l] for the game of figure 5.1(b) rather than the equilibrium [(A,R),r].

An aside is now appropriate on the work of Bernheim (1984) and Pearce (1984). Their rationalizable strategies are constructed by an iterative technique that, in the terminology of this chapter, can be thought of as a candidate for an eductive equilibrating process. It turns out that rationalizing for two-person games consists essentially of the successive delection of dominated strategies as described above. They claim that nothing beyond rationalizing can be adequately justified along properly Bayesian lines. If this were true, even Nash equilibrium for static contests would have to be abandoned. Bernheim comments that "Nash behavior is neither a necessary consequence of rationality, nor a reasonable empirical proposition." In defense of the first of these of assertions, he quotes from Luce and Raiffa (1957, p. 63):

> Even if we were tempted at first to call a [Nash] non-conformist "irrational," we would have to admit that [his opponent] might be "irrational" in which case it would be "rational" for [him] to be "irrational."

The natural reaction is that *of course* Nash equilibrium play should not be anticipated if the opponent might be irrational. This is the reason for the traditional caveat of the implicit axiom of game theory: namely, it is common knowledge that the players are rational.[28] With proper regard to this issue (see the common knowledge references of note 5, particularly Tan and Werlang, 1984), an orthodox Bayesian defense of Nash equilibrium *can* be assembled. At the same time, Bernheim's and Pearce's rationalizability is not free of the type of criticism to be leveled against traditional perceptions of rationality in the next section. This is demonstrated by Reny (1985), whose conclusions on such issues closely parallel some of those offered in this chapter.

None of this is to deny that the rationalizing approach (or the much deeper theory of Kohlberg and Mortens, 1983, 1986) provides insight into games like the dalek example. The point is that it cannot be right, in an eductive context, to treat deviations by player I at x and z as *uncorrelated* (generated by *independent* trembles). The agents[29] I_x and I_z (player I at x and player I at z) have more in common than the fact that they receive the same payoffs. They also share the same model of their environment and hence their predictions of player II's behavior must be related. Thus player II can deduce something from the observation of the play of D by I_x about I_z's prediction of what II will do at y. Very loosely, I would not choose D at x if he believed that II were planning r at y. If I chooses D at x, he must therefore believe that II is planning l at y. In this case, I will play L at z. If so, then it is optimal for II to play l at y and hence for I to play D at x and L at z.

The third principle of Kohlberg and Mertens requires that games like those illustrated in figures 5.1(a) and 5.1(b) be regarded as equivalent. Harsanyi and Selten (1982) deny that this is a compelling rationality requirement. Although this is not a major issue for this chapter, it is as well to note in passing that the approach advocated here would also deny the third principle.[30] For example, in the dalek example of gigure 5.1(b) I do *not* argue that player I will necessarily hold *precisely* the same beliefs about II at x as he holds at z. The argument to be offered against the assumption that players are always able to predict each other's (possibly mixed) choice of action with total precision also entails that players should not be assumed to be able to predict even their *own* future behavior with complete accuracy. Hence the fact that I chose D at x may conceivably serve as useful information to I at z.

5.3 Rosenthal's "centipede" example

A variant of Rosenthal's game is illustrated in figure 5.4. The discussion begins with the simpler game of figure 5.3. This is a game like chess in so far as the outcomes labeled W, D, or L represent a win, draw, or loss respectively for player I. The simpler game is used to make some preliminary points about the doubtful validity of backwards induction and the successive deletion of dominated strategies as never-to-be-denied principles of eductive analysis.

The suffixes refer to the time at which an outcome is achieved. It is assumed that player I's preferences satisfy

$$W_1 > W_2 > \dots W_{101} > D_{50} > L_{52}$$

and that II holds opposing preferences.

A classical analysis, following Zermelo, proceeds by backwards induction (which here is equivalent to the successive deletion of dominated strategies). Working from right to left, nonoptimal actions are successively deleted. In figure 5.3, the undeleted action at each node has been indicated by doubling the edge that represents it. The pure strategies for the two players specified by the doubled edges constitute a unique subgame-perfect equilibrium for the game.

It is not disputed that the result of the play of this game by rational players will be that I plays "down" at the first node. What is to be contested is a statement that is often made about subgame-perfect equilibrium strategies

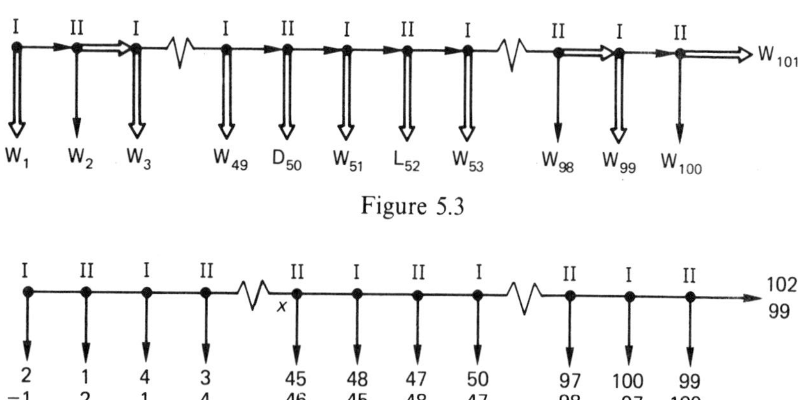

Figure 5.3

Figure 5.4

like those illustrated in figure 5.3. It is said that these represent rational plans of action for the players *under all contingencies*. The inference is that, *if* player II were to find herself called upon for a decision at node 50, then she *would* play "down." But this is a very dubious proposition. Admittedly, it makes good sense if player II explains her arrival at node 50 with some version of the Selten trembling-hand hypothesis. But is this a good explanation? Should she really attribute her arrival at node 50 to 25 uncorrelated random errors on the part of player I or should she look for some less unlikely explanation?

It is important to emphasize that such a question cannot be answered *in the abstract*. The answer depends on the environment in which the game is played. To insist on one answer rather than another is therefore to make a judgment about the nature of this environment. My own judgment about the environment in which chess is normally played would lead me to attribute player I's repeated decision to use a dominated action to some *systematic* error on his part, perhaps relating to some misunderstanding about the rules or payoffs. There would then be good reason to play "across" at node 50 in the hope that this systematic error would be repeated by I at node 51, thus allowing victory for II at node 52. In the same way, good poker players do not play maximin strategies against those they have good reason to suppose to be poor players. They deviate from the maximin strategy in the hope of exploiting the expected bad play of their opponents.

This interpretive difficulty concerning subgame-perfect equilibria can be traced to the logic implicit in the Zermelo algorithm. As Harsanyi and Selten (1980, ch. 1, p. 22) make clear in distinguishing between a "material implication" and a "subjunctive conditional," these logical questions are not entirely transparent.

Let $P(n)$ mean "rational play reaches node n," $Q(n)$ means "rational play at node n is as in figure 5.3," and $R(n)$ mean "$P(n)$ implies $Q(n)$." First it should be shown that $R(n)$ implies $R(n-1)$. Assume $R(n)$ and $P(n-1)$ are true but $Q(n-1)$ is false. Then $P(n)$ follows from $P(n-1)$ and so $Q(n)$ is true. But this is a contradiction. For example, when $n = 50$, player II can achieve a draw by playing "down" but, if rationality requires her to play "across," the argument says that she will lose at node 51. Thus $R(n)$ implies $R(n-1)$. But $R(100)$ is true because $Q(100)$ is necessarily true. Thus $R(1)$ is true by (backwards) induction. Since $P(1)$ is vacuously true, $Q(1)$ follows. But note that the truth of $Q(1)$ *refutes* $P(50)$. Thus, $R(50)$ tells us *nothing* about what a rational player should do if node 50 were *actually* reached because $P(50)$ is *false*. In fact, matters are worse. The suppose deduction of $Q(50)$ from $R(51)$ and $P(50)$ is spurious. If it were correct, then $P(51)$ would be *false* and hence $Q(51)$ could not be deduced from $R(51)$.

The convoluted logic underlying this argument is visited again later in this section. The immediate point is that there is nothing intrinsically irrational or illogical in II's deviating from her subgame-perfect strategy at node 50. Whether she does so will depend on how she models her environment. If a Selten trembling-hand model is appropriate, then players will not deviate from their subgame-perfect strategies that will then represent "rational plans of action under all contingencies." However, for more complex environmental models, this will not necessarily be true and, in the eductive context, it seems reasonable to suggest that more complex models are appropriate.[31]

Rosenthal's (1981) "centipede' example of figure 5.4 is used to highlight the unease felt about the finitely repeated prisoner's dilemma and similar situations. Zermelo's algorithm works as in the preceding example. If the last node is reached, a rational player II will choose "down." If the penultimate node is reached, a rational player I who predicts the choice of "down" should the last node be reached will also choose "down." This argument applies at every node and the ostensible conclusion is that rational play requires I to play "down" at the first node. That such a conclusion makes sense for an *eductive* analysis is now to be denied. *Evolutive* arguments both in favor of the outcome of Zermelo's algorithm and against are not hard to construct. In particular, it is easy to see how the supposedly rational outcome might be generated by a particularly myopic adjustment process. But it is *eductive* processes with which this section is concerned.

What beliefs are to be attributed to a rational player II at node x? I claim that she does not even have to believe that player I and herself have played irrationally in reaching node x in order that the Zermelo outcome be destabilized. All that is necessary is that there be some fuzziness about what is the right way to play. Such a view makes it possible for rational players to delay the play of "down" until late in the game in the manner exemplified at the end of this section. No apology need be made for introducing the idea that unresolved doubt may exist about the "right" way to play. As argued in section 5.6, imprecisions are *necessarily* intrinsic to any properly based theory of rational behavior. In essence, *all* rationality is imperfect. Thus, an imperfect rationality approach, as advocated by Selten (1978) for the closely related "Chain-store paradox" cannot be evaded for problems of this sort.[32]

This observation forces a return to the logic underlying the argument used earlier in undermining the conventional interpretation of a subgame-perfect equilibrium. *Counterfactuals* are involved. Lewis (1976) offers as an example: "If kangaroos had no tails they would topple over." The analysis of the meaning of this sentence given by Selten and Leopold (1982) is particularly apt for what follows. They suggest that such sentences make no sense except in the presence of a background theory.[33] As an implicitly understood background theory, they propose the "mechanical statics of animal organisms." Given such a theory, a computer model of a kangaroo could then be constructed. The parameters of the model could then be varied so as to remove the tail. It would then be possible to test the stability of the resulting construct using computer simulation.

Conventional arguments on subgame-perfect equilibria require counterfactuals of the form: "If a rational player made the following sequence of irrational moves, then ...." To make sense of this along the lines suggested by Selten and Leopold, it is necessary to have a background theory of a rational player that is sufficiently well specified to allow a computer model to be constructed and sufficiently flexible to admit variation of its parameters. The first consideration necessitates an algorithmic (or constructive) view of rationality rather than an ideal (or axiomatic) view. The second necessitates serious consideration of theories of imperfect rationality even by those who are unwilling to be persuaded that perfect rationality is an unattainable ideal.

What is important is that the meaning of the counterfactual "Rational players made the following sequence of irrational moves" depends on the background theory and also, which is a point that Selten and Leopold seem to neglect, on the particular parameter changes employed to "remove the kangaroo's tail" when several such changes are possible. One hesitates to follow Lewis (1976) in proposing a grand scheme involving minimal deviations in the space of all possible worlds, but it is clear that Occam's razor is at least relevant. In particular, a parameter variation that attaches high probability to the observed facts ought to take priority over one that attaches low probability to them.

From this point of view, parameter variations that attribute deviations from rationality to a trembling hand (i.e. to *uncorrelated* random errors) will be an explanation of last resort. Instead, explanations will be sought in which a simple alteration in the structure of a player-model explains many deviations *simultaneously*. If the model is complex, as it will be in an eductive context, this will often be possible and hence errors will have to be seen as *correlated*.

This may clear the air for a more specific analysis of the centipede example. Numerous authors have reached similar conclusions about similar games (e.g. Selten, 1978; Kreps and Wilson, 1982b; Basu, 1985; Reny, 1985). All these share a basic intuition about the root of the difficulty but differ on how this is expressed. My contribution here is simply to express this commonly held intuition within the particular framework of this paper.

Consider first the centipede with all but the last three nodes deleted (so that node 98 becomes the first node). The resulting game is then just a little more complex than the two-node games usually used in justifying subgame-perfect equilibria. The intention is to defend the occasional use of "across" in this truncated version of the centipede.

What conclusion does player I draw from the observation that node 99 has been reached? If he believes that correct play for II at node 98 is "down," he then has the problem of explaining a deviation from correct play.

Depending on how he models himself and his opponent, he may explain this deviation in numerous ways by varying the parameters in these models. A good Bayesian will then attach subjective probabilities to the different explanations. For simplicity, this treatment follows Kreps and Wilson (1982b) in looking at only one particularly simple explanation, although it will be obvious that the argument can be carried through in much the same way with less naive hypotheses. Probability $\varepsilon > 0$ is assigned to the event that player II's reasoning is sufficiently defective that she always plays "across," but with probability $1 - \varepsilon$ her reasoning is not defective at all. If I believes that correct reasoning at node 98 requires playing "down," then arrival at node 99 will signal to I that II is a defective reasoner. Thus I will play across at node 99. But then nondefective reasoning by II will lead to the play of "across" at node 98 and "down" at node 100. Thus "down" at node 98 cannot be correct under these circumstances. Similarly, "across" cannot be correct. Instead, I should play "across" at node 98 with probability $p = 2\varepsilon/(1 - \varepsilon)$. The total probability that a player II (defective or nondefective) will play "across" at node 98 is then 3ε, which exceeds ε.

Thus, even in the truncated centipede, the play of "across" at node 98 by a nondefective II can be explained. In the untruncated centipede, the same argument can.be employed but to greater effect, since the probabilities have a chance to accumulate. However, a looser defense is also available based on the observation that, in similar games after similar histories, players should normally be expected to choose similar (possibly mixed) actions. In the centipede, it is only necessary to observe that, at nodes 20 through 30, say, the players do indeed face similar subgames after similar histories.

5.4 Selten's "horse" example

The preceding example is intended to show that, by discarding all strategy profiles that are not Nash equilibria, conventional game theory sometimes throws out too much. A less heterodox view is that more still should be thrown out. Nothing in this chapter is intended to deny the latter proposition for a wide variety of games (e.g. static games). An example used by Selten (1975) and Kreps and Wilson (1982a) that is designed to illustrate the inadequacy of Nash equilibrium is given in figures 5.5(a) and 5.5(b). Since it is a game with no subgames, it also illustrates the inadequacy of subgame-perfect equilibrium.

The game has two Nash equilibria in pure strategies that will be labeled A and B. These are indicated by doubling the edges that represent equilibrium choices. In fact, there are two classes of Nash equilibria of which A and B are representatives. The first has I and II playing as in A and II playing r with probability $p \geq \frac{3}{4}$. The second class has I and III playing as in B

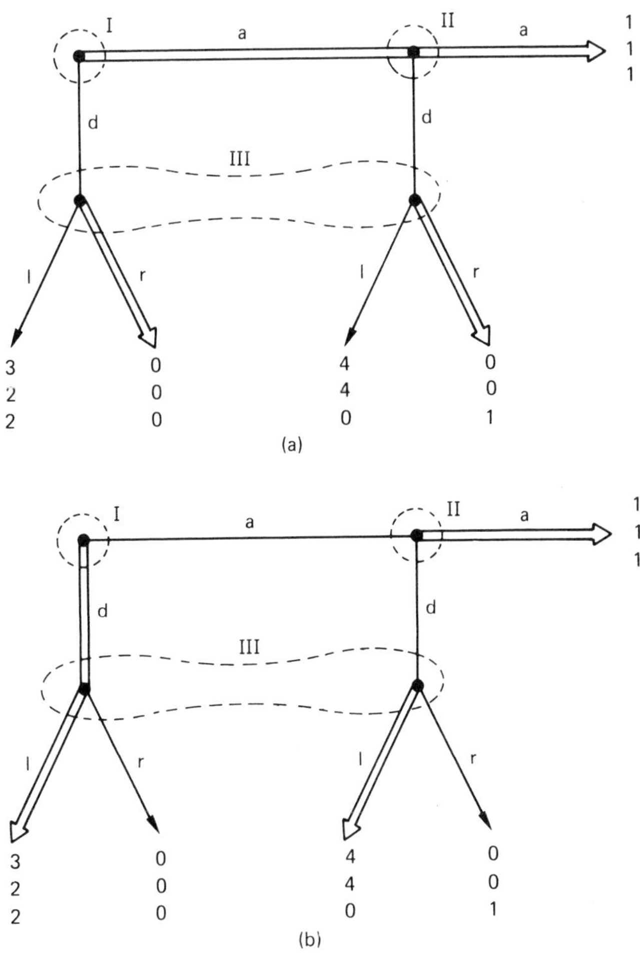

Figure 5.5

and II playing d with probability $q \leqslant \frac{2}{3}$. Type A equilibria all yield the outcome vector (1,1,1). Type B equilibria all yield (3,2,2). The former are therefore Pareto dominated by the latter.

In spite of this observation, the consensus is to reject equilibria of type B because they are not sequential equilibria (and hence not trembling-hand perfect). The essential point is that II's planned behavior in a type B equilibrium is regarded as unreasonable on the grounds that, *if* II were to find herself called upon to make a decision, she would *not* choose action a *given that* III is to choose action l.

The purpose of this section is to deny that this consensus view is necessarily valid for an *eductive* analysis of the game given the sharp attitudes to rationality that accompany this view. Since a central theme of this chapter is that such sharp attitudes to rationality are inappropriate, one is therefore free to treat the proposed eductive argument as a *reductio ad absurdum.* However, it will not be denied that the rejection of type B equilibria makes good sense in an *evolutive* context. If I, II, and III are drawn at random from populations of whom nearly all play type B strategies nearly all the time, then an observer can deduce, from the fact that II has been called upon to play, that I must have played exceptionally. But such an observer will have *no reason to revise her prediction of what III will play.* The observer will therefore regard action d as optimal for player II. Evolutionary forces will therefore tend to remove those player IIs who employ action a. Type B equilibria will therefore not be viable. Indeed, in this evolutive setting, it is tempting to gild the lily by eliminating all but the type A equilibrium in which III plays r with probability $\frac{3}{4}$. If I and II always play a, then no evolutionary pressure constrains III in the direction of r. One would therefore expect an erosion in the number of IIIs playing r and a tendency for the relative fractions playing l and r to equalize (as in the phenomenon biologists call genetic drift). This drift will continue unchecked until slowed and finally reversed by switching behavior on the part of the other players when the probability of playing r falls below $\frac{3}{4}$.

After this defense of the rejection of type B equilibria in an evolutive context (the context of greatest relevance to positive economics), an eductive argument in favor of the retention of type B equilibria is now offered.

The essential hypotheses are that rational thought processes will necessarily bring different players arguing from the same premises to precisely the same conclusions always, and that the proposition that the opponents are rational is to be sustained to the last possible defensive line. Moreover, all this is commonly known.[34]

Since I and III will make the *same* prediction about what II would do if called upon to make a decision, II cannot dismiss I's preceding action is irrelevant to what III will do *if* II is actually required to act (unless, perhaps, II cannot reconcile I's choice of action a with rational behavior). Player I's action provides information about his a priori prediction of II's play and hence information about III's a priori prediction of II's play.[35]

Not only II knows that I and III will make the same a priori prediction of II's planned play. Both I and III know this also. Suppose that it is common knowledge that both predict that II plans to play a with probability $1-q$ and d with probability q. Then I and III will see themselves faced with a two-person game whose normal form is given in figure 5.6. Recalling that this is a *reductio*, traditional arguments may be invoked for the use of a Nash equilibrium in this game.

The strategy pair (a,r) is *always* an equilibrium of the game. It is the *only* equilibrium unless $0 \leqslant q \leqslant \frac{2}{3}$, in which case (d,l) is also an equilibrium together

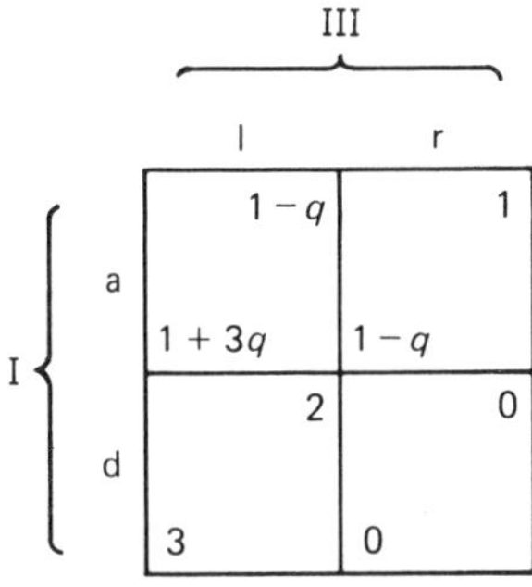

Figure 5.6

with a mixed equilibrium in which III plays r with probability at most $\frac{2}{3}$. So what should II deduce from the observation that I has chosen action a?

Certainly, she cannot deduce that I and III have used the Nash equilibrium (d,l) in figure 5.6, since then I would not have played a. Nor can she consistently hold that I and III used the mixed Nash equilibrium. If this were the correct deduction, then she would respond by choosing d. But I and III would predict this and hence make $q = 1$. However, the mixed equilibrium only exists when $0 \leqslant q \leqslant \frac{2}{3}$. Thus, unless II abandons the hypothesis that I and III have chosen rationally, she is left with the Nash equilibrium (a,r) for figure 5.6. Her optimal response to these choices by I and II is to play a.

Of course, I and III will reason all this through and hence make $q = 0$. Observe that the fact that II will attribute the observation of the choice of a by I to the selection of Nash equilibrium (a,r) in figure 5.6 does not constrain I and III in their *actual* choice of a Nash equilibrium. In particular, nothing excludes their choice of (d,l). Indeed, this is particularly attractive since it Pareto dominates the alternatives. But, if players I and III choose d and l respectively and player II chooses a, then their joint actions generate the equilibrium (d,a,l). This is the type B equilibrium rejected by traditional arguments.

If the defense of (d,a,l) offered above is accepted, the player II begins with the belief that the play of a by I is impossible. On observing such an impossible event, she recalculates. Rather than surrender the hypothesis that player I has behaved rationally, she reassesses her estimate of what rational play consists of. One might say that she is subject to trembles in what she supposes to be the correct way to play. Such a tremble, however, is very different from those envisaged in a trembling-hand equilibrium.

The purpose of the above argument is *not* to claim that (d,a,l) is the *right* equilibrium in an eductive context. Its purpose is simply to expose the existence of factors that are ignored in the usual approach. A convincing eductive analysis would need a very much more precise specification of the model of a rational player to be used.

5.5 Rationality

In section 5.1 the importance of taking on board not only the substantive aspects of rationality but also its procedural aspects was emphasized. Such an imperative makes it inevitable that an attempt be made to model a rational player as a suitably programmed computing machine. Gödelian problems of self-reference then need to be confronted directly. The fact that no way round these problems exists seems to indicate that perfect rationality is best thought of as unattainable ideal. However, just as mathematicians see no reason to abandon formal deductive arguments in consequence of the failure of the Hilbert program, so decision theorists should see no reason to abandon appeals to rationality simply because perfection may not be fully reachable.

In this section, some of the difficulties will be explained in a game-theoretic context. These difficulties are nothing more than the Liar's paradox and Newcomb's paradox dressed up in mathematical language.[36]

The basic idea that needs to be borrowed from mathematics is that of a Turing machine. Essentially, this is a computing device with no predetermined upper bound on the amount of storage it may use in calculation. Turing envisaged a machine with a *finite* number of internal states equipped with a device for reading or writing letters from a fixed alphabet on a paper tape of indefinite length. Before the calculation, the tape is left blank except for the input data. After the calculation, it is to be hoped that the tape will contain an answer to the problem the machine is designed to resolve. During the calculation, the tape may be used to store interim results. What happens at any step in the calculation depends on the internal state of the machine and the symbol on the tape at the beginning of the step. These factors determine what the machine does with the tape *and* the next internal state to which the machine moves. The machine may overwrite symbols on the tape and/or scroll the tape one space to the left or right.

Such a machine is quite a primitive object and it might be thought surprising that it is orthodox for mathematicians to subscribe to the Church–Turing thesis, which asserts that any formal calculation possible for a human mathematician can be aped by a Turing machine. It is possible that this orthodoxy is mistaken, but no counterexample is known. In any case, no apology would seem to be necessary for modeling a player as a Turing machine.[37,38]

Since a Turing machine is a finite object, its design can be specified completely using a finite list of symbols. The set of all possible designs is therefore enumerable. Each Turing machine may therefore be uniquely associated with a natural number, which will be referred to as its Gödel number. Turing demonstrated the existence of *universal* Turing machines. If such a machine is offered the Gödel number x of *any* other machine on its tape followed by a piece of data d, then the universal machine is able to mimic the action of machine x on that piece of data, i.e. it will print the same output as machine x would print given data d.

Universal Turing machines are particularly relevant for game theory because of its traditional implicit assumption that perfectly rational players can duplicate the reasoning process of their opponents and hence predict accurately the mixed strategy that an opponent will use. Universal Turing machines certainly can duplicate the activity of any other machine. Moreover, the proviso about adequate data would seem covered by the "implicit axiom" mentioned in section 5.1: namely, it is "common knowledge" that all players are "perfectly rational."

But matters are not quite so comfortable. A player might begin with the belief that "I am perfectly rational and so are the other players: therefore they reason just like me." However, it has been argued earlier that this is not a view that can sensibly be sustained if the unfolding of the game yields systematic deviations by the opponents from the predictions generated by this belief. To deal with this situation in an *eductive* context, a player needs to be able to cope with hypotheses about the reasoning processes of the opponents other than simply that which maintains that they are the same as his own. Any other view risks relegating rational players to the role of the "unlucky" bridge expert who usually loses but explains that his play is "correct" and would have led to his winning if only the opponents had played "correctly." Crudely, rational behavior should include the capacity to exploit bad play by the opponents.

In any case, if Turing machines are used to model the players, it is possible to suppose that the play of a game is prefixed by an exchange of the players' Gödel numbers. Each player would then be supplied with the information necessary for an eductive analysis to be possible. Some such preliminary exchange of information about Gödel numbers, the implicit communication mentioned in section 5.1, would seem to be intrinsic to an eductive approach (although this is not to argue that an eductive approach necessarily requires *full* disclosure of Gödel numbers).

Within this framework, a perfectly rational machine ought presumably to be able to predict the behavior of the opposing machines perfectly, since it will be familiar with every detail of their design. And a universal Turing machine *can* do this. What it *cannot* do is to predict its opponents' behavior perfectly *and* simultaneously participate in the action of the game. It is in this sense that the claim that perfect rationality is an unattainable ideal is to be understood.

The following is a defense of the above proposition. Consider a two-person game to be played by Turing machines. Before the game is played, each machine receives as input all relevant data in the form

$$([g], [h], [x], [y])$$

Here $[g]$ encodes the rules of the game and the role allotted in the game to the machine receiving the data while $[y]$ encodes a complete description of the machine occupying the opposing role. The opposing machine receives the same data in the different order

$$([h], [g], [y], [x])$$

when an analogous interpretation. The square brackets are meant to indicate that the data has been expressed in a form suitable for computer ingestion.

As output each machine is expected to supply an analysis of the game in the form

$$([p], [q])$$

where q is a list of recommendations about how to play the game when filling the role assigned to the analyzing machine and p is a list of predictions about the recommendations of the other machine on the basis of which q can be evaluated. A list of tautologies will not be acceptable as a prediction, i.e. a prediction is to be understood as something capable of being falsified by the opponent's recommendation. Of course, a machine might product no output at all, or produce an output that is not in the specified form.[39] In either case, the machine will fail to supply an analysis.

One may prefer to think of the machines as providing advice for human players rather than as the players themselves. The output of a machine can then be viewed in much the same manner as a book on chess. Such a book contains not only advice on what to do under various contingencies, it also explains why this advice is thought to be good by analyzing what would happen if the advice were not followed. In particular, this requires making predictions about how a good opponent will respond to deviations.

Because universal Turing machines exist, we can find a machine z which, when offered the signal $([g], [h], [x], [y])$, mimics the action of machine y on the data $([h], [g], [y], [x])$. If y would produce the analysis $([p], [q])$, then z prints $([P], [Q])$, in which P is chosen to be a *true* prediction about q and Q is chosen to be a recommendation that *falsifies* the prediction p. If y would produce an output which is not acceptable as an analysis, then z prints a prespecified analysis $([P_0], [Q_0])$. If y would calculate forever, without ever producing an output, then z will necessarily do the same.

The case of interest is when $x = z$ and y is a putatively perfectly rational machine r. If the latter is *always* required to produce an output, which may perhaps be TILT if it does not like that data it is given, then r either will fail to make a prediction of the recommendation Q_0 or else will make a prediction

p that is false of the recommendation Q. Neither outcome is very satisfactory for a supposedly perfectly rational machine.

The proviso that r must actually finish its calculation is important in the above. In general, a Turing machine may calculate forever without reaching any conclusion. Moreover, there is no effective procedure that will determine, for any given machine, whether that machine will eventually come to a half after receiving a given piece of data. As an example of a Turing machine calculation that never terminates, it is only necessary to take $g = h$ (making the game symmetric) and $x = y = z$. The given argument then leads to a contradiction since $P = p$ and $Q = q$. But P is a true list of statements about q, and p is not a true list of statements about Q. Hence these outputs are not produced. This simply formalizes the commonplace observation that "If I think that you think that I think ..." involves an infinite regress.

These considerations isolate the difficulty with the notion of a perfectly rational machine. Sometimes, such a machine, in seeking the perfect answer to a question, would wish to calculate forever. But a machine that calculates forever without providing an answer is not a very useful machine. Two routes out of this difficulty suggest themselves. That advocated in part II of this study is to restrict attention to those machines that incorporate a device that *guarantees* that a conclusion is reached for *all* relevant inputs. The crudest such device would allot a fixed time to a calculation. If the time limit is exceeded without a conclusion being reached, then the machine would abandon the attempt to predict the opponent's analysis precisely and use some 'guessing algorithm" instead. The important point is that the insistence that machines *always* produce an output means that their predictions will necessarily sometimes be in error regardless of the sophistication of the stopping advice.

The traditional route out of the difficulty would be to rule out from consideration those machines x for which r does not supply a valid analysis when offered the data $([h], [g], [r], [x])$. It is certainly true that the awkward machine z constructed above is not an optimizer[40] and so grounds can be found for excluding it. Indeed, in so far as traditional discussions can be reinterpreted in Turing machine language, they would seem to go much further by excluding all machines x except r itself.

The "unlucky expert" argument has already been cited against such an approach. A more fundamental difficulty lies in the nonconstructive nature of the technique. A definition that requires that a perfectly rational machine only be perfectly rational when playing a perfectly rational machine is circular. But if there is a perfectly rational machine, then there should be a perfectly rational method for checking that it is indeed perfectly rational (see Binmore, 1984).

Even if such a nonconstructive definition is admitted, it does not follow that Gödelian problems can necessarily be evaded. Instead of an awkward opponent, awkward games can be proposed (provided one is allowed to define these nonconstructively also). Suppose that r is a perfectly rational machine, however defined. Next suppose that g is a machine that produces payoffs for 2×2 symmetric matrix games as follows. First it prints payoffs that make pure strategy 1 strictly optimal for both players. Then it mimics the operation of machine r when offered data $[g]$. Either machine r will halt or it will not. If it does not halt, then the originally printed payoffs stand. If it does halt and proposes pure strategy 1, then machine g erases the payoffs it has printed and replaces them by payoffs which make pure strategy 2 strictly optimal. If it halts and proposes a (possibly mixed) strategy which is not pure strategy 1, then machine g does nothing. Two perfectly rational machines playing a game whose payoffs are determined by g will give the *wrong* answer. The problem is well posed in the sense that there exists another machine R that will output the optimal strategy for the Catch-22 game g but otherwise mimics r. Such a machine R would be more perfect than perfect! (The definition of the game is nonconstructive because of the lack of an effective procedure – a procedure guaranteed to give an answer in finite time – for deciding whether Turing machines will or will not halt.)

None of this is at all profound. Mathematically, all that is involved is a trivial adaptation of the standard argument for the halting problem for Turing machines.[41] Indeed, many mathematicians would regard an argument as unnecessary, seeing the conclusions as an essential consequence of Gödel's observation that consistency is incompatible with completeness. In summary, the claim is that, if attention is restricted to players who *always* give an answer to problems that make proper sense, then sometimes such players will get the answer *wrong*. It is not claimed that the case for this proposition is watertight: only that those who deny it need to be prepared to explain why they should not be classified along with those who count angels on the end of pins.

5.6 Bayesianism

How does all this relate to the prevailing Bayesian orthodoxy? In this section, I propose to argue briefly that the standard contemporary interpretation of Savage's (1954) *Foundations of statics*, in which he synthesized Von Neumann and Morgenstern's expected utility theory with the subjective probability ideas of Ramsay, De Finetti, and others, is *naive*. Savage's theory is entirely and exclusively a *consistency* theory. It has nothing to say about *how* decision-makers come to have the beliefs ascribed to them; it asserts only that, if the decisions taken are consistent (in a sense made precise by a list of formal axioms), then they act *as though* they maximize expected utility relative to a subjective probability distribution. Objections to the axiom system can be made, although it is no objection in an eductive context to argue that real people often contravene the axioms. People also often get their sums wrong, but this is no basis for proposing a change in the axiomatic foundations of arithmetic. In any case, it is not the consistency axioms which are to be criticized here.

What is denied is that Savage's passive descriptive theory can be reinterpreted as an active prescriptive theory at negligible cost. Obviously, a reasonable decision-maker will wish to avoid inconsistencies. A naive Bayesian therefore assumes that it is enough to assign prior subjective probability distributions to a decision-maker and then to forget about the problem of where beliefs come from. Consistency forces any new data that may appear to be incorporated into the system by Bayesian updating, i.e. a posterior distribution is obtained from the prior distribution using Bayes's rule (hence Harsanyi's coining of the term *Bayesian*). The naiveté does not consist in using Bayes's rule, which is a trivial manipulation whose validity as a piece of algebra is not in question. It lies in supposing that the prior can be chosen from a limited stock of standard distributions without much, if any, in the way of soul-searching.

Savage's view was considerably more complex. He did indeed argue that his descriptive theory could be of assistance in helping decision-makers to form their beliefs. His point was that reasonable decision-makers would not rest if they found inconsistencies in their belief systems. What then should they do? To quote Luce and Raiffa (1957, p. 302), expounding Savage's view:

> Once confronted with such inconsistencies, one should, so the argument goes, modify one's initial decisions [about beliefs] so as to be consistent. Let us assume that this jockeying – making snap judgments, checking on their consistency, modifying them, again checking on consistency etc. – leads ultimately to a bona fide, a priori distribution.

It may be helpful to expand on this quotation. A person who makes judgments in a sensible way will presumably prefer to make judgments when he or she has more information rather than less. A decision-maker might therefore begin to address the problem of constructing a suitable belief system by asking: for *every* conceivable possible course of future events, what would my beliefs be *after* experiencing them? Such an approach automatically discounts the impact that new knowledge will have on the basic model being used to determine beliefs, i.e. it eliminates the possibility of being *surprised* by an event whose implications have not previously been considered. Next comes the question: is this system of *contingent* beliefs consistent? If not, then the decision-maker may examine the relative confidence he or she has in the judgments that have been made and then adjust the corresponding beliefs until they *are* consistent. But consistency, in this context, is equivalent to asserting that the system of contingent beliefs can be deduced, using Bayes's rule, from a common prior. It is therefore true, in this story, that the final "massaged" posteriors can be deduced from the final "massaged" prior using Bayes's rule. This is guaranteed by the use of a complex adjustment process which operates *until* consistency is achieved. As far as the massaged beliefs are concerned, Bayes's rule therefore has the status of a *tautology*, like $2+2=4$.

Sophisticated Bayesians sometimes signal their recognition that some analog of Luce and Raiffa's jockeying procedure is required by asserting that Bayesian theory applies only to "closed universe" problems, i.e. to problems in which all potential surprises *can* be discounted in advance. Suitable examples are to be found in the small finite decision trees with which books on Bayesian decision theory for students of business administration are illustrated. One aim of the preceding section was to make it clear that it is far from obvious that game-theoretic problems can be treated as closed universe problems. The necessity of using a model that incorporates entities that are as complex as the internal structure of the decision-maker guarantees otherwise.[42]

Naive Bayesians make no such qualifications. For them, all universes are closed and all calculations complete. What they overlook is that the Luce and Raiffa jockeying story tells us *nothing whatever* about how scientific inferences should be made. The story sweeps the problem of scientific inference under the carpet by absorbing the relevant issues into the *wholly unspecified* adjustment process by means of which primitive snap judgments are massaged into a consistent belief system. Once the problm is under the carpet, in the sense of not appearing explicitly in the mathematical formalism, it can then be conveniently forgotten altogether. Very naive Bayesians are then even able to ask whether there is any point in game theory at all. The argument is that Bayesian players must have a subjective probability distribution over the strategy choices of their opponents. In a static game, all that therefore remains for a Bayesian player is to choose his or her own strategy so as to maximize expected utility relative to this subjective distribution. No particular reason then exists for supposing that the resulting strategy choices will constitute even a Nash equilibrium.

In this very naive argument, the issue of where beliefs come from has been completely overlooked. Naive Bayesian rationality apparently endows its fortunate adherents with the capacity to pluck their beliefs from the air. But this will not do for game theory. One might almost say that what game theory is *about* is the massaging process (via "If I think that you think ..." arguments) by means of which beliefs are constructed.

More sophisticated Bayesians impose restrictions on the beliefs that their players pluck from the air which are based on the requirement that the players' beliefs ought to be mutually consistent, in some sense. Exactly how this is done depends on what is assumed to be common knowledge (see note 5). This static approach has its virtues for the analysis of static games but the theme of this paper has been that an approach to equilibria that seeks to evade the process by means of which equilibrium is attained leads to intractable difficulties for games with some dynamic structure.

Harsanyi and Selten (1980, 1982) have made a serious attempt to describe a suitable equilibrating process (incorporating much more than just their tracing procedure). I only wish to comment here that, although they express themselves in Bayesian language, their process incorporates features that are not traditionally thought of as Bayesian. Indeed, some of the features have been criticized as *ad hoc* and arbitrary: unfairly, if this paper is to be believed, since the unattainability of perfection implies that any approach will necessarily be arbitrary to some extent.

5.7 Conclusion

In this chapter some aspects of current game theory have been criticized. Its thrust is that an attempt must be made to model players' thinking processes explicitly. The most demanding reason is so that deviations from predicted play can be "explained" by modifying the model once this has proved inconsistent with observed events. Chapter 6 seeks to clarify some of the issues relevant to such a modeling attempt in an eductive context.

NOTES

1 Extensive-form games of imperfect information.
2 Selten (1975) offers subgame-perfect equilibria and trembling-hand perfect equilibria. Kreps and Wilson (1982a) have sequential equilibria. Kohlberg and Mertens (1983) have a notion usually referred to as "stability," more recently (1986) refined to "hyperstability." Myerson (1978) has proper equilibria. Kalai and Samet (1982) have persistent equilibria. Banks and Sobel (1985) outbid Selten with divine equilibria. There are also the notions of Cho and Kreps (1985) and, with a somewhat different focus, the work of Bernheim (1984) and Pearce (1984). Finally, there is the Harsanyi and Selten tracing procedure (Harsanyi, 1975).
3 Consider, for example, *existence*. This is regarded as a *sine qua non* for an equilibrium notion by those brought up in the Bourbaki tradition. But evolutionary stable equilibria do not always exist. Is the idea therefore to be abandoned? Clearly not. The nonexistence of such an equilibrium simply signals that the dynamics of the equilibrating process are likely to be sufficiently wild that unmodeled constraints will be rendered active.
4 Hayek (1948, p. 91) makes an odd bedfellow on this issue.
5 Of course, not all analyses can be criticized in this way. Examples of exceptions

are the tracing procedure of Harsanyi and Selten (Harsanyi, 1975) and the "rationalizing algorithm" of Bernheim (1984) and Pearce (1984).

6 Only recently have formal attempts been made to deal properly with the role of common knowledge assumptions in the foundations of the subject. See, for example, Aumann (1974, 1983), Brandenburger and Dekel (1985a, b, c, 1986), Reny (1985) and Tan and Werlang (1984, 1985). Also, see Werlang (1986).

7 As, more recently, Bernheim (1984) has hijacked the irreplaceable word *rationalize.*

8 Of sufficient complexity.

9 It is tempting to suggest that conclusive arguments may not be available in this area. But no argument in support of this suggestion could then be conclusive.

10 Perhaps the word "libration" might be used instead of the clumsy "equilibrating process" or the equally clumsy "adjustment process." *Tâtonnement* seems to carry the connotation that the dynamics are only metaphorical.

11 Fudenberg, Kreps, and Levine (1986) make a similar point in a narrower but very much more precise context.

12 Even less is it a good explanation for one's own past bad play. Introspection seldom brings me to this conclusion.

13 Which is not to deny that animals are more clever (and humans more stupid) than it was once customary to suppose.

14 "Every humour ... hath its proper eductive cathartic" (*Oxford English Dictionary*).

15 Although this need not be true of the *behavior* elicited by an appropriate trigger response.

16 Why then study it? The excuse for this paper is that it is commonplace for theorists (e.g. in rational expectations economics) to argue as though from an eductive standpoint. However, as regards positive economics, it would seem to be evolutive processes that matter most.

17 Even in noneductive environments, concern would seem appropriate about stability in the face of *all* trembles. One would expect stability only in respect of trembles actually experienced.

18 Because of Gödelian problems of self-reference.

19 Sometimes game theory is attacked because it is said to ignore ethical considerations. But in a game theory, as envisaged by Von Neumann and Morgenstern, the ends are inseparable from the means. Thus, if the players care that an outcome is achieved via a broken agreement, then the fact that an agreement has or has not been made must be explicitly modeled so that the players' preferences can be properly reflected in the utilities assigned to the game outcomes.

20 For games of complete information, it is usually held that the rules of the game and the tastes and beliefs of the players must be common knowledge. The latter means that not only do all players know the information, but all players know that all players know and all players know that all players know that all players know, and so on. Harsanyi's (1967–8) theory of games of incomplete information treats situations in which there are uncertainties about the tastes and beliefs of the other players by transforming the game into a larger game of complete information but with more players (usually called *types*). This theory does *not* evade the requirement that there be a background pool of common knowledge

of some sort. As long as beliefs, in the above, refer only to formal chance moves in the rules of the game, this paper offers no criticism. It is when over-sharp beliefs are attributed to players about what other players will do in the game that the paper's criticisms become operative.

21 The formulation of Kreps and Wilson (1982a) meets this criterion if an initial chance move is appended to the structure.

22 Note that I is required to state his planned choice at z even though his planned choice at x may ensure that z is not reached.

23 "Correlated equilibria" involve strategies that are conditioned on unformalized events ("sunspots") from which the players derive correlated information. The existence of such correlated information disbars the analysis of the game as a contest in the sense of this chapter. It can be made into a contest by formalizing the unformalized events which the players use to coordinate their strategy choices.

24 Even this is questionable as a general principle. Static 2×2 matrix games exist that have a *unique* Nash equilibrium that assigns each player only his or her security level. Why do players therefore not deviate to their security level strategies and hence *guarantee* these payoffs?

25 Although Kohlberg and Mertens (1983, 1986) would perhaps disagree.

26 Originally Selten used the term "perfect" but decided (1975) to reassign the word to what I later call a "trembling-hand perfect" equilibrium.

27 It is unnecessary to consider subgame-perfect equilibria because all subgames in the perturbed game will be reached with positive probability. This implies that Nash equilibria are necessarily subgame-perfect. (Note that the trembling-hand equilibrium attained will depend on the manner in which the perturbations become negligible.)

28 And, *of course*, Nash equilibrium will not be a reasonable empirical proposition in the laboratory unless this requirement is satisfied to some extent – perhaps through repeated play against anonymous opponents drawn from a fixed pool. The evidence cited against Von Neumann's maximin theory of two-person zero-sum games, for example, is almost entirely worthless because of its neglect of this and other obvious preconditions of the theory. For more extensive comments, see Binmore (1987).

29 Selten (1975) splits his players into "agents," each of whom controls one and only one information set. These agents act independently although, of course, all the agents acting for a single player have the same preferences as that player. (Formally, trembling-hand equilibria are the limits of Nash equilibria in perturbed versions of such "agent normal forms.")

30 But note that Harsanyi and Selten's reason for arguing that sequential agent splitting may be significant differs from that offered here.

31 Subgame-perfection is built into the Kreps–Wilson idea of a sequential equilibrium. In agreeing that weaker conditions on their belief system may sometimes be appropriate, they warn against weakening the conditions too much because of the risk of losing subgame-perfection (1982a, p. 876). (The difficulties considered above are unconnected with their formal belief system. The beliefs attributed to player II cannot be expressed within their formalism.)

32 Although not necessarily with Selten's specific model of rationality, which is

more convincing as a model of *homo sapiens* than as a model of *homo economicus.*

33 As a meaningless counterfactual, they offer: "If leaves were blue, foxes would be green." The point is that no background theory exists.

34 My own view is that these hypotheses do not even make proper sense because they treat rationality in a sense which involves internal inconsistencies.

35 This requires a denial of the Kreps and Wilson observation that "defections from the equilibrium strategy ought to be uncorrelated" (1982a, p. 875).

36 Just as the difficulty with backwards induction discussed in section 5.3 was "nothing more" than that faced by the condemned prisoner who is told that he will be executed next week on a day when he is not expecting it and wrongly concludes that he will not be executed at all.

37 Simon's (1955, 1959, 1977) notion of "bounded rationality" would seem to involve modeling players not as Turing machines but as finite automata. However, the important distinction is that Simon is concerned with players whose internal complexity is low compared with that of their environment, whereas an eductive analysis requires admitting players whose internal complexity may be very large indeed. When thinking about bounded rationality, the finite automata considered are therefore usually small in size.

38 Sometimes concern is expressed over the fact that Turing machines, being finite, cannot handle, for example, real numbers. But this is to misunderstand what human mathematicians can do. A mathematican can do no more in specifying a number than to write down a finite list of symbols (a fact embodied in the Löwenheim–Skolem theorem which asserts that all mathematical systems have a countable model). In any case, Gödelian problems do not disappear if machines with an infinite number of states are permitted.

39 The question of whether an output is or is not a valid analysis is to be understood as effectively decidable.

40 Indeed, the example does not specify payoffs at all and so optimization is undefined.

41 Other results from mathematical logic and computation theory can be employed to the same end as, for example, in McAfee (1984). His appeal to Blum's "speed-up" theorem is particularly pleasing.

42 In fact, there is a whole body of mathematics that cries for attention in this context. The axiom of choice, in Zermelo–Fraenkel set theory, may be loosely interpreted as asserting the existence of abstract computing devices (functions) whose internal structure is beyond the capacity of the analyzing mathematician to duplicate. He or she is therefore faced with an "open universe" problem. It is clearly no accident that, with the axiom of choice, sets of real numbers, which are not Lebesgue measurable, exist but, without it (but with some ancillary technical assumptions), *all* sets of real numbers can be taken to be measurable. De Finetti (1974) must be given credit for being more scrupulous than some in that he acknowledges the related Banach–Tarski paradox. But his response, that only *finite* sets be admitted into the universe of discourse, misses the point. The infinite, in this context, serves as an idealization for nonconstructible. This problem will not go away and nothing is served by adopting a formalism within which it cannot be expressed.

REFERENCES

Abreu, D., and Rubinstein, A. 1986: "The structure of Nash-equilibrium in repeated games with finite automata," ICERD Discussion Paper 86/141, London School of Economics.

Aumann, R. 1974: "Subjective and correlation in randomized mixed strategy," *Journal of Mathematical Economics* 1, 67–96.

Aumann, R. 1983: "Correlated equilibrium as an expression of Bayesian rationality," mimeo, Hebrew University, Jerusalem.

Banks, J., and Sobel, J. 1985: "Equilibrium selection in signalling games," mimeo, MIT, Cambridge, Mass.

Bass, K. 1985: "Strategic irrationality in extensive games," mimeo, Institute for Advanced Study, Princeton, N.J.

Bernheim, D. 1984: "Rationalizable strategic behavior," *Econometrica* 52, 1007–28.

Binmore, K. G. 1984: "Equilibria in extensive games," *Economic Journal* 95, 51–9.

Binmore, K. G. 1987: "Experimental economics," *European Economic Review* 31, 257–64.

Brandenburger, A., and Dekel, E. 1985a: "Common knowledge with probability 1," Research Paper 796R, Stanford University, Calif.

Brandenburger, A., and Dekel, E. 1985b: "Hierarchies of beliefs and common knowledge," Research Paper 841. Stanford University, Calif.

Brandenburger, A., and Dekel, E. 1985c: "Rationalizability and correlated equilibrium," mimeo, Harvard University, Mass.

Brandenburger, A., and Dekel, E. 1986: "Bayesian rationality in games," mimeo, Harvard University, Mass.

Cho, I., and Kreps, D. 1985: "More signalling games and stable equilibria," mimeo, Stanford University, Calif.

De Finetti, B. 1974: *Theory of Probability*. New York: Wiley.

Friedman, J., and Rosenthal, R. 1984: "A positive approach to non-cooperative games," mimeo, Virginia Polytechnic Institute and State University, Blacksburg, VA.

Fudenberg, D., Kreps, D., and Levine, D. 1986: "On the robustness of equilibrium refinements," mimeo, Stanford University, Calif.

Harsanyi, J. 1967: "Games of incomplete information played by Bayesian players," parts I, II and III, *Management Science* 14, 159–82, 320–34, 486–502.

Harsanyi, J. 1975: "The tracing procedure," *International Journal of Game Theory* 5, 61–94.

Harsanyi, J., and Selten, R. 1980: "A non-cooperative solution concept with cooperative applications," chapter 1, draft, Center for Research in Management, Berkeley, Calif.

Harsanyi, J., and Selten, R. 1982: "A general theory of equilibrium selection games," chapter 3, draft, Bielefeld Working Paper 1114, Bielefeld.

Harsanyi, J., and Selten, R. 1988: *A General Theory of Equilibrium Selection in Games*. Cambridge, Mass.: MIT Press.

Hayek, F. von 1948: "Economics and knowledge, in *Individual and Economic Order*. Chicago, Ill.: University of Chicago Press.

Kalai, E., and Samet, D. 1982: "Persistent equilibrium in strategic games," Discussion Paper, Northwestern University, Ill.

Kline, M. 1980: *Mathematics, the Loss of Certainty*. Oxford: Oxford University Press.

Kohlberg, E., and Mertens, J. 1983: "On the strategic stability of equilibria," Operations Research and Economics, Discussion Paper 8248, Université Catholique de Louvain.

Kohlberg, E., and Mertens, J. 1986: "On the strategic stability of equilibria," *Econometrica* 54, 1003–37.

Kolmogorov, A. 1950: *Foundations of the Theory of Probability*. New York: Chelsea.

Kreps, D., and Wilson, R. 1982a: "Sequential equilibria," *Econometrica* 50, 863–94.

Kreps, D., and Wilson, R. 1982b: "Reputations and imperfect information," *Journal of Economic Theory* 27, 253–79.

Lakatos, I. 1976: *Proofs and Refutations, the Logic of Mathematical Discovery*. Cambridge: Cambridge University Press.

Lewis, D. 1976: *Counterfactuals*. Oxford: Basil Blackwell.

Luce, R., and Raiffa, H. 1957: *Games and Decisions*. New York: Wiley.

Maynard Smith, J. 1982: *Evolution and the Theory of games*. Cambridge: Cambridge University Press.

Marschak, T., and Selten, R. 1978: "Restabilizing responses, inertia supergames and oligopolistic equilibria," *Quarterly Journal of Economics* 92, 1–93.

McAfee, P. 1984: "Effective computability in economic decisions," mimeo, University of Western Ontario.

Moulin, H. 1981: *Théorie des jeux pour l'économie et la politique*. Paris: Hermann. (A revised version in English is published by New York University Press under the title *Game Theory for the Social Sciences*.)

Myerson, R. 1978: "Refinements of the Nash equilibrium concept," *International Journal of Game Theory* 7, 73–80.

Myerson, R. 1984: "An introduction to game theory," Discussion Paper 623, Northwestern University, Ill.

Myerson, R. 1986: "Credible negotiation statements and coherent plans," Discussion Paper 691, Northwestern Universith Ill.

Nash, J. 1951: "Non-cooperative games." *Annals of Mathematics* 54, 86–95.

Neyman, A. 1985: "Bounded complexity justifies cooperation in the finitely repeated prisoners' dilemma," *Economics Letters* 19, 227–9.

Pearce, D. 1984: "Rationalizable strategic behavior and the problem of perfection," *Econometrica* 52, 1029–50.

Reny, P. 1985: "Rationality, common knowledge and the theory of games," mimeo, Princeton University, N.J.

Rosenthal, R. 1981: "Games of perfect information, predatory pricing and the chain-store paradox," *Journal of Economic Theory* 25, 92–100.

Rubinstein, A. 1985: Finite automata play the repeated prisoners' dilemma," ST/ICERD Discussion Paper 85/109, London School of Economics.

Savage, L. 1954: *Foundations of Statistics*. New York: Wiley.

Selten, R. 1975: "Re-examination of the perfectness concept for equilibrium in extensive games," *International Journal of Game Theory*, 4, 22–5.

Selten, R. 1978: "Chain-store paradox," *Theory and decision* 9, 127–90.

Selten, R. 1983: "Evolutionary stability in extensive 2-person games," Bielefeld Working Papers 121 and 122, Bielefeld.

Selten, R., and Leopold, U. 1982: "Subjunctive conditionals in decision theory and game theory," in *Philosophy of Economics*, vol. 2, *Studies in Economics*, eds Stegmuller, Balzer, and Spohn. Berlin: Springer-Verlag.

Sen, A. 1976: "Rational fools," in *Scientific Models of Man*, The Herbert Spencer Lectures, ed. H. Harris. Oxford: Oxford University Press.

Simon, H. 1955: "A behavioral model of rational choice," *Quarterly Journal of Economics* 69, 99–118.

Simon, H. 1959: "Theories of decision-making in economics," *American Economic Review* 49, 253–83.

Simon, H. 1976: "From substantive to procedural rationality," in *Method and Appraisal in Economics*, ed. S. Latsis. Cambridge: Cambridge University Press.

Simon, H. 1977: *Models of Discovery*. Dordrecht: Reidel.

Tan, T., and Werlang, S. 1984: "The Bayesian foundations of rationalizable strategic behavior and Nash equilibrium behavior," mimeo, Princeton University, N.J.

Tan, T., and Werlang, S. 1985: "On Aumann's notion of common knowledge: an alternative approach," Serie B-028-jun/85, Instituto de Matematica pura e aplicada, Rio de Janeiro.

Werlang, S. 1986: "Common knowledge and the game theory," Ph.D. thesis, Princeton University, N.J.

6 Modeling rational players: part II

... why may we not say that all Automata ... have an artificiall life.

Hobbes, *Leviathan*

Even in self-consciousness, the I ... remains a riddle to itself.

Schopenhauer

6.1 Introduction

This is the second part of a two-part study. It can be read independently of the first part provided that the reader is prepared to go along with the unorthodox views on game theory which were advanced in part I and are summarized below. The body of the chapter is an attempt to study some of the positive implications of such a viewpoint. This requires an exploration of what is involved in modeling "rational players" as computing machines.

The basic tenet borrowed from part I is the claim that traditional game theory is unsatisfactory in so far as the behavior of ideal "perfectly rational" players is treated axiomatically à la Bourbaki. In the first place, it does not "deliver the goods." For games of any complexity, confusion reigns supreme about what the "correct" analysis ought to be. Simultaneously, there are simple games for which it is quite clear what the "correct" traditional analysis is but for which the conclusions of this analysis are profoundly uncomfortable. In the second place, doubts are appropriate about the logical foundations of the traditional approach. It is, by now, a cliché that completeness is incompatible with consistency for formal deductive systems.[1] Any such system is therefore "imperfect" in the sense that it can be replaced by a better system. Traditional theory seeks to evade this issue by confining its attention exclusively to *substantive* questions (what do players do?) and entirely neglecting *procedural* questions (how do players decide what to do?). As Simon (1976) has observed, this is a criticism that is relevant not only to game theory but to economic theory in general.[2] In game theory, however, matters are particularly serious. Indeed, in the first part of this study I argued

---

This chapter, and its predecessor, which consist of amplified and revised extracts from a working paper (1986) of the same title, were written with the support of NSF grant number SES-8605025. I am indebted to numerous individuals for much useful comment, but particularly to Ariel Rubinstein.

that it is *because* traditional theory neglects procedural questions that it has failed to "deliver the goods."

To seek to tackle procedural questions seriously is to commit oneself to an attempt to model the thinking processes of the players *explicitly*. Traditional game theory lacks such a model and hence is helpless in the face of the counterfactual: suppose a perfectly rational player made the following sequence of irrational moves, then .... But an equilibrium analysis cannot evade such counterfactuals. What keeps players on the equilibrium path is their expectation of what *would* happen if they *were* to deviate from equilibrium play. However, the traditional straitjacket makes it very hard to confront such issues squarely. The result is the construction of magnificent mathematical edifices of which a medieval scholastic might justly be proud, but little in the way of genuine progress.

The existence of an explicit model of a player allows out-of-equilibrium play to be treated in a nonmetaphysical manner. To illustrate this point, a lighthearted example is borrowed from Selten and Leopold (1982) who, in turn, borrowed it from Lewis (1976). Consider the counterfactual: if kangaroos had no tails, they would topple over. Selten and Leopold argue[3] that, to make sense of such a counterfactual, it is necessary to have available a background theory which is adequate to admit the construction of a computer *model* of a kangaroo. The parameters of such a model may then be varied so as to "remove the kangaroo's tail." The stability of the resulting construct can then be tested (perhaps by simulation), thereby attaching a meaning to the superficially nonsensical counterfactual.

Thus, in game theory, the observation of an unanticipated[4] sequence of moves on the part of an opponent should lead to an updating of the model used to describe his or her thinking processes and hence to a *revision of the predictions of his or her future behavior*. Of course, in some situations such a procedure will result in "hustling"[5] and other attempts at deceit. But untangling such attempts is part of what game theory should be about.

Such a viewpoint forces attention to be paid to issues which traditional game theory leaves unformalized and therefore, in accordance with the Bourbaki ethos, neglects altogether. In particular, the *process* by means of which equilibrium is reached (the "libration") will necessarily determine the *nature* of the equilibrium achieved. Part I of this study distinguished two types of equilibrating environment: the *evolutive* and the *eductive*. The former is undoubtedly the more significant for positive economics. In such an environment, the players are seen as simple stimulus–response machines whose behavior has the appearance of having adapted to the behavior of other machines because ill-adapted machines have been weeded out by some form of evolutionary competition. The dynamics of the libration process are therefore *external* to the players and visible over time to an observer as a game is played repeatedly.

This chapter, however, follows most economic theorists in confining attention to the *eductive* context.[6] This means that equilibrium is achieved through careful reasoning by the agents before and during the play of the game. Sometimes the use of equilibrium strategies in this context is made part of the axiomatic characterization of rationality. But this evades the fundamental question: namely, what is the "right" equilibrium? To resolve this question, it is necessary to reduce the problem to an appropriate one-person problem in which players deduce information about the expected play of their opponents through some form of introspection. Central to this problem are the reasoning chains which begin, "If I think that you think that I think ...." Such considerations are *internal* to a player. The dynamics of an eductive libration are therefore invisible to an observer. But the point of view we are pushing here insists that this is no good reason for proceeding as though the dynamics were absent altogether.[7]

The use of computing machines (automata) to model players in an *evolutive* context is presumably uncontroversial. For evolutive work, the emphasis belongs on machines of *low* complexity compared with their environment, in accordance with Simon's (1955, 1959, 1977) notion of bounded rationality. In part I of this study, it was argued that computing machines are also appropriate for modeling players in an *eductive* context. Mathematicians are sold on the idea that they can model themselves, in their aspect as formal calculators, in this way.[8] One can therefore defend the use of computing machines, in an eductive context, by claiming that players are to be modeled as mathematicians of the formalist school. Such computing machines, however, must be understood to have the potential to be of very *high* complexity compared with their environment. In particular, there is no reason why such machines should not reprogram themselves as the game proceeds.

Further comment on the previous paragraph is postponed until section 6.2. The point which requires emphasis at this stage is that, even in an eductive context, with a machine that has access to *all* relevant information and which can compute for an *arbitrarily* long period, it is *necessarily* the case that the machine will sometimes be in error in the predictions it makes about its opponents. The reason is essentially that the machine would sometimes calculate forever if this were permitted. To avoid this, a "stopping rule" must be built in. If such a stopping rule guillotines an exact calculation, then the machine will be forced to employ an alternative "guessing algorithm." By its nature, such a guessing algorithm will sometimes guess wrongly. Part I of this study offered a formal argument to this effort, along with the conclusion that "perfect rationality" is therefore an unattainable ideal useful only for metaphysical purposes.[9]

Observe that there is *necessarily* an *arbitrary* element in the choice of stopping-rule-cum-guessing algorithm, if only because it will always be better to stop later and hence guess less frequently. This may be uncomfortable for the mathematically minded, but it has advantages when it comes to "removing the kangaroo's tail'' in order to explain an unanticipated event. From the point of view being advocated here, a rational opponent cannot be seen as a unique "type." There are an infinite number of types of model which can justifiably be described as "rational." This leaves open the possibility of explaining deviations from predicted play without the necessity of abandoning the hypothesis that the opponent is rational. He or she may not be of a type to which high probability was attached originally, but such probabilities can be updated as the game proceeds.

I am aware that the "trembling-hand" explanation of deviations also does not require abandoning the hypothesis that the opponent is rational after a deviation. However, in part I of this study it was argued at length that, *in an eductive context*, the trembling-hand explanation should be one of last resort. This is not to deny that irrational "mistakes" may frequently be needed to explain deviations. All that is claimed is that, when the "kangaroo's tail" can be removed in various ways, unlikely parameter variations of the model should not be ranked above likely parameter variations.

Part II of this study is no more able to offer sharp answers to the problems raised than part I. Since sharp answers would provide an incidental resolution of the problem of scientific induction, this is perhaps not surprising. However, diffuse answers are not without value and those offered do seem to provide some insight into a number of questions. The interpretation of mixed strategies (section 6.4) and the issue of "forwards induction" (the examples of section 6.6 and 6.7) deserve explicit mention. On the latter question, a summary of the general approach advocated for the eductive context is postponed until the conclusion (section 6.8). It may seem perverse to delay this account until the very end, but I am anxious that what is essentially a program for further study not be confused with yet another "equilibrium concept." Perhaps it may provide a framework for an equilibrium concept if a good theory of eductive libration processes ever emerges, and perhaps the considerations of sections 6.2, 6.3, and 6.5 may have a place in such a theory. As things stand, however, possibly their most useful role is in providing a warning that schemes like that offered in the conclusion should not be interpreted too naively.

Finally, as in so many things, it should be noted that Harsanyi and Selten have visited most of these issues before. We refer, in particular, to Selten (1975) on "mistakes," Selten (1978) on "machine models," Marschak and Selten (1978) on "correlated trembles," Harsanyi (1967–8) on "types," and Harsanyi (1975) and Harsanyi and Selten (1980, 1982) on "the tracing procedure."

6.2 Costly rationality

This section is an attempt to relate the ideas offered in this to the more general issues usually discussed under the somewhat misleading heading of "bounded rationality." Megiddo (1986) gives a thought-provoking introduction to these questions, which are both conceptually difficult and technically demanding. However, as always in this chapter, this section confines itself to generalities.

What would a genuinely applicable theory of a rational player be like? Obviously, it would have to take account not only of what is technically feasible in the way of calculation but also of the *costs* of calculating. As Megiddo (1986) and others emphasize, the cost of the *time* used in calculating will be a major consideration. Real-life games always involve some explicit or implicit constraint on the time that may be taken to make a move.[10] A chess clock is perhaps the most evident embodiment of such a time constraint. Accepting this view forces a simultaneous acceptance of the requirement that a model of a rational player will have to incorporate *heuristics*, which it uses to decide whether or not to carry out a specific calculation and to guess, or "value," the results of calculations which it decides not to carry out in detail. (Recall the stopping-rule-cum-guessing algorithm of section 6.1.) This idea will be familiar to those who know something of the mechanics of chess-playing computer programs.

An important incidental is that an attempt to model the heuristic decision-making process along Bayesian lines is hopelessly inappropriate. The point will not be pressed here since it follows from what was said in part I, section 5.6. Briefly, the achievement of consistency cannot realistically be regarded as costless. On the contrary, the necessary calculations should be expected to dwarf anything that has been discussed hitherto.

Of all the many possible models of costly rationality, which should be chosen? One approach is to hypothesize "meta-players" who design the machines which actually play. Megiddo and Wigderson (1986), Neyman (1985), Rubinstein (1985), and Abreu and Rubinstein (1986) see these "meta-players" as playing a "meta-game"[11] in which a pure strategy is the choice of machine. A Nash equilibrium is then sought for this meta-game. A natural criticism is that this procedure evades the issue by transferring the problem of costly rationality from the players of the game to unmodeled meta-players. But these meta-players have to solve a problem which is even more complex than that faced by the players themselves. However, such a criticism is fair only if the meta-players are granted a real, rather than a metaphorical, existence. Consider, by way of analogy, the status of the "auctioneer" in the Arrow–Debreu model of a market. Nobody believes that such an auctioneer really exists, except in rare circumstances. No individual nor organization actually takes on the highly complex task of calculating the market-clearing prices. This is achieved via an unmodeled *tâtonnement* process for which the auctioneer serves as a simplifying substitute. In a methodologically analogous fashion, meta-players can be seen as a metaphor for an *evolutionary* process. The question of the complexity of the decision-making processes attributed to the meta-players then ceases to be an issue since it is unloaded onto the environment. This is to argue that work of this kind should be classified as *evolutive* in character, along with such work as that reported by Maynard Smith (1982, ch. 5) on the evolutionary stability of animal learning rules.

Before more is said about these evolutionary questions, another point needs to be made. Usually, theoretical realizations of bounded rationality incorporate a fixed, *exogenously determined*, upper bound on some aspect of the complexity of the strategies available to a player (i.e. on the complexity of a machine available to a meta-player), and, from now on, the term "bounded rationality" will be used only in this sense in this study.[12] Such an exogenously determined constraint must be expected to be *active* in equilibrium. For example, in Neyman (1985) or Megiddo and Wigderson (1986), cooperation is obtained as equilibrium behavior by boundedly rational players in games like the repeated prisoner's dilemma, precisely *because* the optimal strategies make the constraint active. To quote Megiddo (1985):

> The underlying idea is that, no matter what the scarce resource is, [meta-] players design machines that waste the entire amount of available resource so that they cannot perform even simple tasks like counting the number of stages [in a repeated game].

It was in this sense that the machines which model the players in an evolutive context were said to be low in complexity relative to their environment.

One may think of bounded rationality, as construed above, as a degenerate case of costly rationality in which costs are zero if the exogenously imposed constraint is not violated but unacceptably large if it is. However, this will obviously not do for an eductive analysis, for which a minimal requirement must surely be that the *marginal* costs of calculation must always be very small. Thus, if a machine fails to carry out certain computational tasks in equilibrium, it will be because the machine has *chosen* not to do so, *not* because it is *unable* to do so without abandoning other computational tasks. In an eductive context, any computational constraints on a machine must therefore be *endogenous* (i.e. self-imposed). This point is made in an encouraging paper by Abreu and Rubinstein (1986).[13] It may clarify matters to observe that bounded rationality, in essense, empowers a meta-player to make *commitments* to stay within a certain strategy set. But Selten's reasons (part I, section 5.2) for rejecting commitment in an eductive context remain valid even if the players are modeled as computing machines.

So far in this section it has been argued that some of the recent work, in which costly or bounded rationality has been imported into a game-theoretic framework, can be seen as exercises in *evolutive* game theory. At the risk of confusing matters hopelessly and completely, I now propose to argue that, even in an *eductive* context, evolution has a central role to play. One might even take the view that, without some evolutionary story in the background, rationality, in the sense required for an eductive analysis, makes little sense. Given several rival models of rationality, how is one to make a choice?[14] The practical man's answer is to let them compete and see which survives. This, for example, is the ethos behind "Dutch book" defenses of Bayesianism.[15] Of course, such a view requires interpreting the word "evolution" in an adequately wide sense. The word invites concentration on biological processes, but it is important to draw attention to social evolution also. Genetics doubtless largely determines the brain's hardware, but its *software* is probably more relevant to the concerns of this chapter. To paraphrase Dawkins (1976): memes (ideas, learning rules, behavioral norms, etc.) are just as much the object of evolutionary pressures as genes, but memes multiply through *imitation* rather than physical replication. In particular, education serves as a meme propagator, and this is one of the reasons for the choice of the word "eductive."

What distinguishes eductive theory from evolutive theory is that, in the former, evolution operates *at one remove*. In evolutive theory, evolutionary processes work *directly* on the strategies in *one specific game*. When costly rationality is taken into account, this means that the evolutionary processes act directly on the rules of behavior which implement these strategies. Such rules of behavior can be thought of as computer programs or as computing machines operated by such programs. In eductive theory, on the other hand, evolutionary processes work on a *master-program* which has the capacity to choose strategies in a *wide variety* of disparate games, some of which *may never have been played before*. This distinction will be reflected in the relative complexity of the decision-making process in the evolutive and the eductive cases respectively.

How are the evolutionary origins attributed to a master-program in the eductive context to be operationalized? More is involved than making the choice of constraints on calculation endogenous and, of course, requiring that the costs of calculation be very small. The device of allowing meta-players to select Nash equilibrium strategies in a machine-choosing meta-game will not suffice in an eductive context. This is not for novel qualitative reasons. It is simply that the inadequaties of the expedient in the evolutive case become more pressing in the eductive case.

Imagine a basic collection of possible master-programs. From this "meme-soup" of possible master-programs, individuals are repeatedly drawn at random to act as players in games similarly drawn at random from a wide menu of possible games. A selection mechanism of some sort operates so that master-programs with engender low payoffs become less frequent in the soup than those which engender high payoffs. Think of a *large* population of hosts (the hardware), each of which may be infected with a master-program (the software), with successful master-programs sometimes displacing less successful master-programs. Interest then centers on the master-programs which survive this evolutionary processes.[16] At least two regulatory mechanisms need to be taken into account in considering what governs the complexity of the surviving master-programs. The first is that punishments must be provided for delayed decisions. The second is that complex machines are more likely to malfunction than simple machines.[17]

In order to say something about the master-programs (or machines or memes) which thrive in this evolutionary competition, it is necessary to look for an appropriate equilibrium. But a Nash equilibrium is not an appropriate equilibrium concept. This is *not* because of the familiar fact that repeated play of a game between the *same* opponents admits a whole range of phenomena inexplicable with Nash equilibrium in a one-shot game[18], for example, coooperation, altruism, revenge, threats, etc. (see Aumann, 1981). Each time a new game is played, the players are *reselected* at random and hence the players cannot use the current game as an opportunity to punish a specific opponent for bad behavior in the past, nor to signal to a *specific* opponent that good behavior would be jointly profitable in the future.[19] Nor is a Nash equilibrium rejected because of an inadequate "informational basis." It is understood that the players are to be supplied with all information allowed by the rules of a game playing it (and this includes the payoffs of the opponents to the extent that this is permitted). Moreover, large numbers of observations of the play of games in the past will allow the master-programs to "learn" how relevant characteristics of master-programs are distributed *overall* in the population as equilibrium is approached. An "explanation" is therefore provided of how certain matters can become "common knowledge" as required in classical game theory (although, admittedly, the explanation is only slightly less skeletal than the traditional appeals to "implicit communication" or a "a common cultural heritage").

One reason for rejecting Nash equilibrium is that, if one master-program (or set of closely related master-programs) wins out in the evolutionary struggle, then, towards the end of the struggle, there will be a high probability that such a master-program *will be playing itself* (or a close relative or relatives) when called upon to play. Given that it is the "welfare" of the master-program, rather than the "welfare" of any particular host or hosts carrying the master-program, that matters for evolutionary success or failure, it is clear that a master-program, playing itself, has a *one-person* decision problem to solve rather than a multi-person problem. A mechanism therefore exists for the evolution of *cooperative behavior* as described, for example, by Axelrod (1984). Biologists refer to this as kin selection. A gene which generates behavior which favors relatives, generates behavior which favors itself, because the gene has a good chance of being replicated in the body of a relative. As far as game theory is concerned, these considerations provide some support for the intuition that, if one equilibrium Pareto dominates another, then the former should be preferred.[20] But why not go further and abandon the equilibrium requirement altogether? This is a popular question which is usually asked in the specific context of the one-shot prisoner's dilemma by those who find the game theory solution paradoxical. Hofstadter (1983) gives an unusually clear version of the position of the "anti-equilibrium school." In essense, if I am playing myself, why do I not just maximize the sum of my payoffs? Game theorists usually shrug off this question,[21] but they are not really entitled to do so, given their traditional view that game theory concerns players who all reason in precisely the *same* "perfectly rational" manner.

From the viewpoint advocated here, however, it is not true that a winner in the evolutionary struggle will *necessarily* be playing itself (or a close relative).[22] A small probability will exist, for example, that an opponent may be a "dinosaur" using an obsolete strategy that has not yet been eliminated. Indeed, if the malfulctioning of programs is allowed for, inferior strategies will never be eliminated. Such malfunctioning will ensure a continual invasion of the population of master-programs by "mutants." A winning master-program will therefore have to incorporate a facility designed to detect and identify such deviants (during the play of each separate game) with a view to exploiting them, where possible. More immediately to the point, however, is the fact that a master program will not win the evolutionary struggle in the long run unless it is, itself, immune to exploitation by mutants. Thus, although mutants (and/or dinosaurs) may be present in relatively small numbers, their presence must be expected to have a very strong influence on the behavior of the winning master-program.

To say that the idea of a Nash equilibrium in a machine-choosing meta-game should be replaced by the idea of an *evolutionary stable equilibrium* (Maynard Smith, 1982) would be a gross simplification. For one thing, the nature of the mutants around will be a function of the equilibrium machine. Nevertheless, it does convey the flavor of what is being proposed.

In summary, the use of the device of a machine-choosing meta-game in an eductive context requires that

1 marginal costs of complexity be always small, so that bounds on the amount of computation are determined endogenously,
2 machines be capable of playing any role in any one of a wide variety

of games, including games of which they have little or no previous experience,

3 equilibrium in the meta-game be interpreted in an evolutionary sense (as a minimum, the equilibrium should be immune to invasion by mutant machines created by malfunctions in the machines currently present in the equilibrium population), and

4 machines be capable of recognizing and responding to deviant behavior.[23]

It is, of course, one thing to formulate such dicta and quite another to construct a theory in which they are operationalized. In what follows, their role is merely to provide a standpoint from which to evaluate the structure for a game-playing machine outlined informally in section 6.3 and 6.5.

6.3 Game-playing machines[24]

What follows in this section clearly has wider implications than the game-theoretic applications for which it is proposed. In particular, it has some significance for the software–hardware approach to the mind–body problem advocated by some biologists for the evolution of self-consciousness in humans. However, I have nothing particularly original to offer on these general issues. The aim is to codify what seems to be an emerging consensus on these questions, in a manner that, although far from formal, is sufficiently precise to allow the possible applications in game theory to be sensibly evaluated.

Recall, from part I, that only the play of *contests* is to be considered. I use this term to indicate a game in which no pre-play communication between the players about the game is permitted (nor communication during its play except in so far as the formal moves provided for in the rules of the game can be used for this purpose). To say that a game is a contest is therefore to make a statement about the environment in which the game is played. The study of contests has a good claim to be regarded as fundamental in that games played in environments which do permit some communication can be reduced, in principle, to contests by modeling the communication possibilities as formal moves in a larger game (Nash, 1951; Binmore and Dasgupta, 1987).

The fact that players cannot exchange messages before the play of the game does not imply that they do not share information beyond that supplied to them in accordance with the rules of the game. In the language of the preceding section, they will certainly share the information that they have been chosen from the same population or society. Among other things, this will involve a shared knowledge about arbitrary conventions which will have evolved in the population. Examples are, "Drive on the right," or, "Condition your investment decisions on the level of sunspot activity." Such conventions allow players to coordinate their behavior by making use of the fact that they can commonly observe phenomena which are not intrinsic to the game (such as the labels used to identify strategies or actions).

Sometimes difficulties over such conventions are allowed to confuse rationality discussions.[25] But the issues seem to me to be secondary and best studied by looking directly at the manner in which such conventions actually evolve by *explicitly* modeling the appropriate extrinsic phenomena and employing the theory of repeated games. For this reason, the term "contest" will be taken to imply that the game is played in an environment in which any labeling of events which players observe together has yet to take on a meaning. One may think of this requirement as forbidding the "tacit communication" implicit in a common understanding of the historical significance of arbitrary symbols or extraneous circumstances.

Recall that the basic aim of this paper is to explore what is involved in an eductive libration process. The advantage of "factoring out" communication possibilities that are not explicitly modeled as part of the game (which is what is achieved by looking only at contests) is that pre-play libration takes place entirely *inside the head* of each player *separately*. One consequence is that care is necessary in interpreting the *meaning* of equilibria in an eductive context – particularly equilibria requiring mixed strategies. This point is made at greater length in section 6.4. All that needs to be said at this point is that machines (or master-programs) which survive the evolutionary struggle described in the previous section will exhibit behavior which is adopted to the *mix* of behaviors exhibited by the equilibrium population. Thus, when two "rational" machines face each other in a static game, their choices of strategy will be approximately optimal given their *predictions* of the strategy to be chosen by the other player. But this prediction need not always be realized, either because the equilibrium population may contain many different types of machine (as in Harsanyi's (1967–8) theory of incomplete information), or because the manner in which the machines calculate has some built-in indeterminacy,[26] or both at once. The ideal of "perfect rationality" excludes such possibilities, but the fundamental tenet of this chapter is that this ideal is unattainable.

So far, all that has been achieved is to delimit the class of games to be considered so that all pre-play activity can be located inside each machine separately. Of course, as the game proceeds, players will *learn* about the structure of the machines they are playing against by studying the moves they choose. This will lead them to revise their predictions of what their opponents are likely to do next. Only in static games (i.e. games which can be thought of as taking place at a single instant) do these learning considerations become irrelevant so that only the pre-play calculations matter. For this reason, this section and the next are concerned only with *static* contests. The question of learning about opponents over time is left until section 6.5. This allows attention to be concentrated on the amount of calculation a machine does, although learning issues cannot be neglected altogether.

The basic model I propose is that of a computing machine programmed to write programs to play games. It may therefore be thought of as a "meme-generating meme." For a static game, it receives, as input, a coded description of the game and responds by writing a program which analyzes the game and recommends a choice of (possibly mixed) strategy. An analysis of the game includes a prediction of the behavior of the machines acting as the other players. For simplicity, attention will be confined to two-player games so that only one opponent need be considered. On what basis is a prediction of the opponent's behavior to be made? Note that a description of the opposing machine (its Gödel number) is *not* given as part of a machine's data as in part I, section 5.5, nor are data about the manner in which the *specific* machine currently being played has played games in the past. The possibility of establishing a *reputation* as a certain type of player is therefore absent.

In predicting the opponent's behavior, a machine can therefore only draw on the fact that both itself and the opposing machine have been drawn from the *same* population of potential players. The data available from this source will be classified under two headings: *objective* and *subjective*. The objective data consists of observations of the way games have been played in the past by players drawn at random from the same population. Obviously, data on the way the *current* game, or similar games, has been played will be particularly important. The subjective data consists of the machine's own master-program. (For simplicity, memories of more primitive master-programs that may once have controlled the machine in the past, and may still control other machines in the present, are ignored.) The machine's master-program, it is assumed, is stored like other data and can be accessed to an extent to be discussed later. Notice that the subjective data is just as "hard" as the objective data.

Yet more taxonomy is now required. Two situations will be distinguished in respect of the available objective data. The first occurs when there is a very large quantity of relevant objective data, for example if the strategies chosen by players drawn at random from the population at large have been observed for several million previous plays of the game which is about to be played. The second occurs when the quantity of relevant data is not very large: perhaps only a few previous plays of the current game, or similar games, have been observed (although the requirement that evolution has brought about equilibrium in regard to the master-programs requires that there must be a very large amount of objective data available on *dissimilar* games).

The two situations are distinguished because it seems to be only the second which is of genuine interest for an *eductive* analysis. Unless one wishes to contemplate the survival of master-programs whose predictions are at variance with the empirical evidence, the first situation is essentially *evolutive* in character. A master-program need only write a program to play the game which maximizes expected utility given the empirically observed probability distribution over the opponent's strategies. Complexity problems are therefore unloaded onto the environment. My suspicion is that those who insist that the choice of an equilibrium strategy should be taken as an *axiomatic* requirement of "rationality" actually have this evolutive setting in the back of their minds. In such a setting, the axiom is certainly defensible, provided that the word "equilibrium" is not over-specified. For example, game-playing machines will be "satisficers" to the extent that they will carry through calculations only to an appropriate level of *approximation.* But, as Radner (1980) has pointed out (see also Megiddo, 1986), very different results can sometimes follow when precise equilibria are replaced by approximate equilibria.

It is when the amount of objective data is inadequate for a "naive" empirical approach that an eductive analysis becomes necessary. Any libration dynamics can then no longer be treated as being entirely external to the players and hence requiring no justification on rationality grounds so that equilibrium has to be treated as a fundamental principle. In the situation of interest for an eductive approach, if "rationality" generates equilibria, this fact must be *explained*, together with the reasons for the selection of one equilibrium rather than another when multiple equilibria exist. Somehow, the machine's subjective data must be harnessed to this end, along with whatever available objective data may bear on the matter.

An aside is now necessary on data which may be available at one remove via the reports of others. As far as objective data is concerned, this is no problem provided appropriate reservations are made about the reliability of material obtained secondhand. But what of reports from other machines about *their* subjective data? What of the deductions that can be made about the subjective data of other machines from observing their general game-playing behavior? Both these considerations are important to the assumption that lies at the heart of the approach advocated here: namely, that introspection is a viable method for predicting the behavior of others. More specifically, in order to predict what an opposing machine will do in a situation for which adequate objective data is not available, the assumption will be that a machine will use, as a guide, what it would do itself, if it were in the same situation. Such a procedure makes sense only if the answers to the preceding questions are supportive of such a predictive device. To consider the plausibility of this requirement, it is necessary to return to the putative evolutionary origins of the game-playing machines. Recall that the basic mechanism seen as driving the evolutionary process in section 6.2 was *imitation*, with an emphasis on the role of *education* in facilitating imitation. Thus, as a machine learns about the subjective data present in other machines in the population, it must be expected to incorporate into its own master-program those elements of other master-programs which are successful and to suppress those elements of its own original master-program which are less successful. In the long run, the tendency will be to generate a population with *closely similar* master-programs and hence its members will have good reason to suppose that introspection is a valuable source of information about the thinking processes of others. It is not claimed, of course, that such a conclusion is *necessary*: only that it serves as a useful working hypothesis.

How does a machine make use of its subjective data in seeking to predict the behavior of the opponent?[27] Sometimes, of course, it will not *need* to make such a prediction, as in the *one-shot* prisoners' dilemma. Let us, however, confine attention to those cases where a prediction (albeit an approximate prediction) is *necessary* if the situation is to be reduced to a one-person optimization problem. It will then be assumed that the machine seeks to simulate the reasoning processes of the opponent by running its own master-program[28] with the opponent's input data and using the output as a prediction of the opponent's play.

But this is not so straightforward as it may seem. The evolutionary defense given above for the use of introspection as a tool in predicting the behavior of others relies on machines' having the facility to modify their own programs as they learn from experience. The machines are therefore *self-correcting*. However, if a machine is programmed to correct itself if the prediction of what it would do in some future circumstance fails some criterion, then the prediction will not necessarily be correct because the predicted behavior is subject to correction.[29] A self-correcting machine cannot therefore predict even its own behavior with certainty.[30]

The nuts-and-bolts reason for this difficulty is that a machine which runs *precisely* its own program before finally deciding what action to take is involved in an infinite regress.[31] To simulate a machine that simulates its own operation requires a simulation of a simulation which, in turn requires a simulation of a simulation of a simulation; and so on. If the machine is not to calculate forever, a stopping rule is required to break out of this infinite sequence of nested loops. When this stopping rule calls for a halt, the machine will necessarily have to call upon a "guessing algorithm" in order to predict its own behavior. Given that a guessing-algorithm-cum-stopping rule is required, where does it come from? In particular, what is the origin of the basic guessing rule?

Here the fact has to be faced that an answer to this question must be *arbitrary* to a large extent, unless the evolutionary process offered as an "explanation" of the origin of the game-playing machines is to be modeled explicitly. Even then, the answer will presumably be contingent on such historical *accidents* as the composition of the original population from which the current population evolved. This is not to say that nothing can be said at all. If nothing else, the second law of thermodynamics will tell us something about the probabilities which a guessing rule should attribute to events which are neutral to evolutionary pressures.[32] However, the general problem of scientific induction is beyond the scope of this chapter and no attempt will be made to explore these issues. It will simply be assumed that a machine *does* incorporate an algorithm which is somehow capable of converting whatever paucity of objective data is available into a preliminary prediction of what it would do if it were called upon for a decision in various circumstances. This basic guessing algorithm is employed only when the stopping rule prevents more elaborate introspection.

What now follows on stopping rules is only a little less sketchy. Given a basic guessing rule, the point of iterating simulations is to *refine* whatever estimate of the opponent's behavior is current. Instead of working downwards, through lower and lower levels of simulation, until the basic guessing rule is invoked, one can think of starting with the basic guessing rule and then working upwards. The machine can then be ascribed an inductive structure in which moving one rung up the simulation ladder is seen simply as a device for replacing one prediction by a more refined prediction. The stopping rule then appears as a rule of thumb for determining when convergence has been sufficiently approached,[33] i.e. when the estimated (small) costs of moving one more rung up the ladder outweigh the estimated benefits of a more refined prediction. These estimates, like the basic guessing rule, will necessarily be *arbitrary* to some degree. Thus the stopping rule will also have arbitrary features.

It would be easy to underestimate the complexity implicit in the structure discussed so far. Recall, however, that the machines under discussion are *self-correcting*. This means, in particular, that the stopping-rule-cum-guessing algorithm will be vulnerable of self-correction. Consider first the implications in respect of the stopping rule.

A consequence of what was said earlier about self-correcting machines is that such a machine cannot "know" (i.e. have full access to) all the operating details of those aspects of its operation which it monitors with a view to possible correction.[34] To this extent, a self-correcting machine will therefore necessarily be a "riddle to itself." But it need not be assumed that the machine is "unaware" of the existence of elements of its own operation to which it does not have full access. Such ignorance may be taken account of, when the machine seeks to simulate its own behavior, by its using, not one simulating subroutine, but many alternative simulating subroutines each of whose outputs is then weighted using an appropriate probability[35] to produce a final prediction.

As far as illustrative examples are concerned, the technically easiest way of taking account of such ineradicable uncertainties requires the assumption that the machine's hardware admits a facility for accessing "black boxes" which generate random numbers.[36] The interiors of these "black boxes" (or their programs) are assumed not to be available for inspection, although nothing precludes the making of statistical inferences on the basis of their output. The advantage of working with machines equipped with such randomizers[37] is that it becomes possible to proceed on the assumption that all machines have an *identical* structure even though their *behavior* need not be identical. It must be remembered, however, that such randomizing "black boxes" are *not* an *essential* feature for a game-playing machine. They are a mathematical device for representing certain unavoidable areas of ignorance in a technically tractable fashion.

Figure 6.1 illustrates the structure of a game-playing machine x which seeks to predict an opponent y, given that a random stopping rule is utilized. The notation $G_n(x)$ $(n \geqslant 1)$ denotes a mixed strategy for player x which is

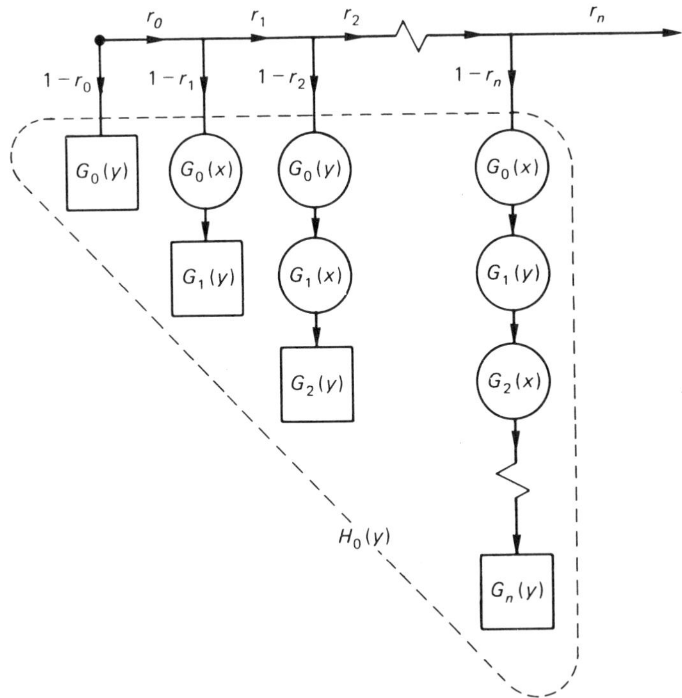

Figure 6.1

an optimal reply to the use of the mixed strategy $G_{n-1}(y)$ by the opposing player y. Some unspecified tie-breaking rule is taken for granted here, but this is not a point which deserves close attention. The basic guessing rule, denoted by G_0, is used with probability $1-r_0$. The nth-level refinement G_n is used with probability $r_0 r_1 \dots r_{n-1}(1-r_n)$.

Consider next the implications of a self-correcting facility for the basic guessing rule. Observe that, in figure 6.1,

$$H_0 = (1-r_0)G_0 + r_0(1-r_1)G_1 + \dots + r_0 r_1 \dots r_{n-1}(1-r_n)G_n \tag{6.1}$$

is supposedly an improvement on the guessing rule G_0. In principle, the machine should therefore be able to improve itself by replacing G_0 throughout by H_0. The new structure may then itself be vulnerable to improvement; and so on. Such self-correction will be worthwhile until the basic guessing rule is no longer much altered by further iterations. If G_0 in figure 6.1 is now assumed to be a basic guessing rule generated by such self-improving "bootstrapping," then it will itself be about as good a predictor as the method is capable of producing. A stopping rule that attaches a high probability to going beyond G_0 would then be computationally inefficient. The implication is that, with a "good" basic guessing rule, the probabilities r_0, r_1, and so on should be very small. But then quadratic terms in (6.1) will be negligible. With a "good" basic guessing rule G_0, figure 6.1 should therefore be replaced by the even simpler structure shown in figure 6.2(a), where r_0 is understood

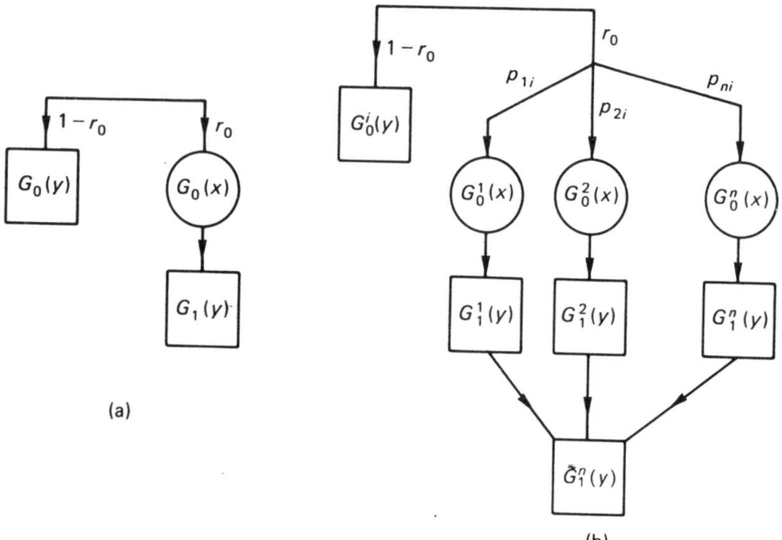

Figure 6.2

to be very small.[38] Of course, the simplicity of figure 6.2(a) is only superficial, being achieved only by concealing the complexities inside the algorithm G_0.

Before leaving figures 6.1 and 6.2(a), some comment on convergence issues may be worthwhile. It will be familiar from the well-known "cobweb model" that the type of bootstrap iteration procedure described here may oscillate violently if predictions for the next period are made dependent only on behavior in the current period (see, for example, Binmore, 1983, pp. 339, 389). On the other hand, stability is achieved if a suitable weighted average of past behaviors is used for predictive purposes. The current section provides some sort of theoretical backing for such a procedure. However, no results are offered on convergence beyond the computations embodied for the specific examples illustrated in figures 6.3 and 6.4.

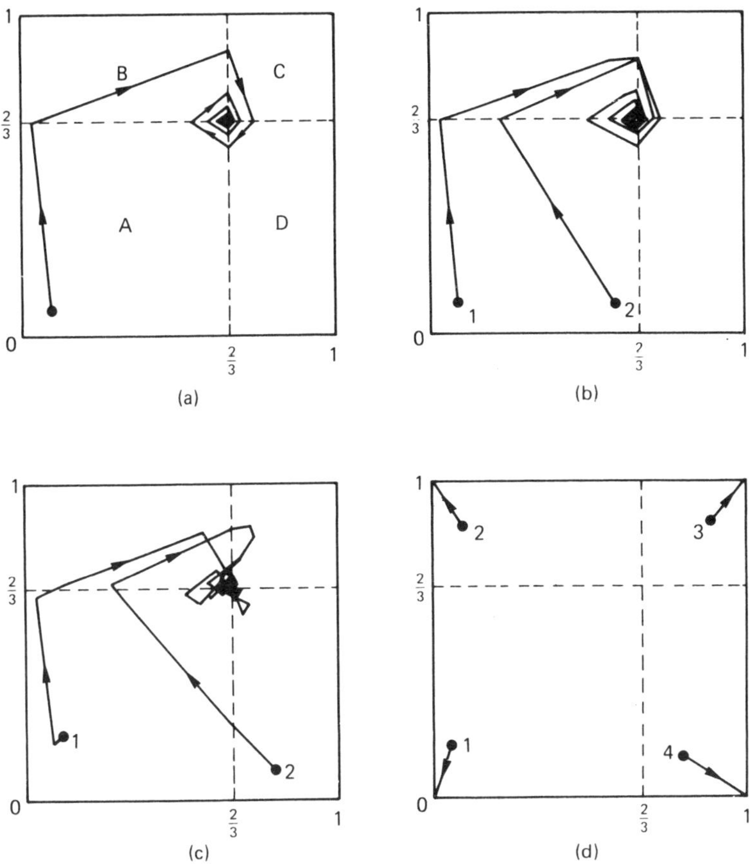

Figure 6.3

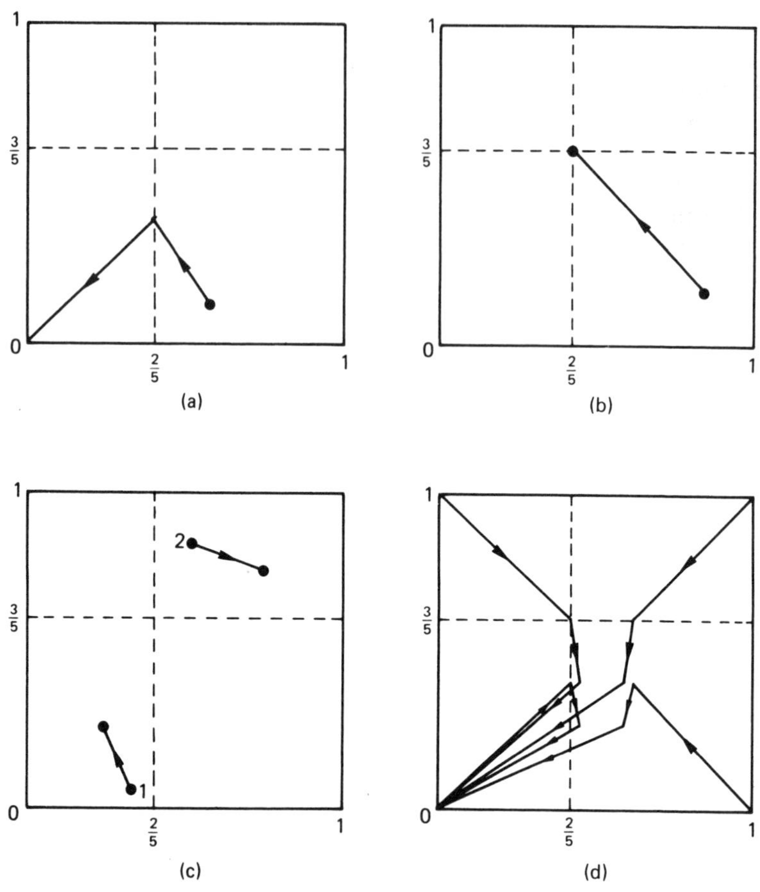

Figure 6.4

In summary, in this section it has been argued that a game-playing machine in an eductive context needs to be seen as a *complicated* device in which loops are nested within loops to an indefinite degree. Not only this, their structure will be *arbitrary* to some extent, and there will be aspects of even its own operation to which a machine will not have full access. Moreover, such *ignorance* is ineradicable. But this is not to argue that eductive game theory is a hopeless enterprise. On the contrary, the bootstrap iteration argument offered here provides a useful supplement to the more traditional arguments employed when attention is focused on Nash equilibria in static contests (see figures 6.3(a) and 6.4(a) for examples). Finally, it should be noted that, although the details of what is proposed differ markedly, the general approach to which the argument leads closely resembles the *tracing procedure* of Harsanyi and Selten (1980, 1982).

6.4 Mixed strategies

The improvement of the basic guessing rule G_0 obtained by replacing G_0 by the rule $H_0 = (1 - r_0)G_0 + r_0 G_1$ (see figure 6.2a) will cease to be worthwhile when H_0 is nearly the same as G_0. (This does *not* necessarily imply that G_0 is nearly the same as G_1, because $G_1 - G_0 = (H_0 - G_0)/r_0$ and r_0 will be small for the reasons given in the preceding section.) Figure 6.3(a) illustrates the dynamics for the zero-sum game with the payoff matrix of figure 6.5.

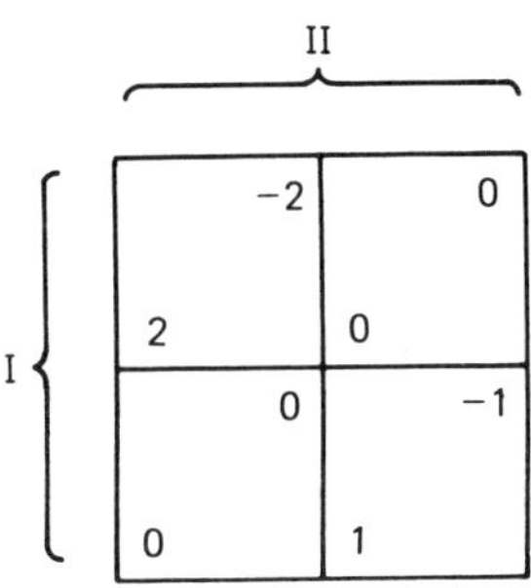

Figure 6.5

A point (p, q) in figure 6.3(a) is to be identified with a basic guessing rule G_0 in which $G_0(\mathrm{I}) = (1 - p, p)$ and $G_0(\mathrm{II}) = (1 - q, q)$. The trajectory marked with arrows indicates the path traced if an initial basic guessing rule originally located at the point (0.1, 0.1) is successively replaced by a guessing rule $H_0 = (1 - r_0)G_0 + r_0 G_1$, when r_0 is vanishingly small. (For example, the optimal reply rule G_1 to an initial basic guessing rule G_0 corresponding to the point (p, q) located in region A of figure 6.3(a) corresponds to the point (0, 1). Thus H_0 will correspond to that point on the line segment joinint (p, q) and (0, 1) which is at a distance from (p, q) equal to a fraction r_0 of the total length of the line segment. Next H_0 is placed in the role of G_0 and the calculation is iterated. Eventually, the boundary between regions A and B is reached, whereupon the optimal reply rule G_1 jumps to (1, 1) and so a new line segment has to be considered.)

Such trajectories are perhaps best thought of as providing a test for a "good" basic guessing rule. Thus the basic guessing rule G_0 corresponding (0, 1/2) fails the test because a refinement of G_0 leads along the trajectory to something significantly different from G_0. Only basic guessing rules close to the unique Nash equilibrium of the game at (2/3, 2/3) will pass the test, in this particular instance.

However, the fact that all trajectories in this game converge on its unique Nash equilibrium is not the point here. Any other result would be surprising.

What is at issue in this section is the *interpretation* of a Nash equilibrium when mixed strategies are involved.

In this zero-sum game, if player I *randomizes* between his pure strategies, using the first with probability 1/3 and the second with probability 2/3, then any strategy is an optimal response for player II, including that in which she *randomizes* between her pure strategies using the first with probability 1/3 and the second with probability 2/3. Since the same is true with the roles of I and II reversed, the quoted pair of *mixed strategies* is a Nash equilibrium for the game. This manner of presenting the idea forces the question: *why* should rational players randomize? They will necessarily be *indifferent* between all pure strategies to which the recommended mixed strategy attaches a positive probability.

The answer is, like much else in game theory, that the wrong question has been asked. The argument which led us to the unique Nash equilibrium in the zero-sum game of figure 6.3(a) was an argument about the convergence of *guessing rules*. The actual *choice* of strategy made by a player is another matter.[39] This will depend on the stopping rule which controls the bootstrap iteration by means of which the basic guessing rule is refined. If, for example, this happens to leave the final pair of predictions in region D of figure 6.3(a), then an optimal response by both players is to use their first pure strategy (i.e. $p = q = 0$). They will therefore make this choice, although the predicted return on using the first pure strategy will only be a bit better than that for the second pure strategy. The point here is that they do not *choose* to randomize, i.e. the output from a machine will *not* normally designate a mixed strategy.[40] But this output will be predictable by the other machine only to the extent that the guessing rule allows. This conclusion is a consequence of the observation that certain aspects of its own operation must necessarily be hidden from a self-correcting machine. This is particularly easy to understand when random stopping rules are employed, as in the illustrative example of section 6.3 used to motivate the dynamics of this section. But the same conclusion *necessarily* applies even when attention is confined to machines whose entire operation is *totaly deterministic*.

The supposed paradox – that game theory necessarily requires rational players to randomize – is therefore a chimera.[41] When a Nash equilibrium calls for the use of mixed strategies, an eductive analysis requires that the probabilities involved be seen as reflecting only the ineradicable uncertainties that a player will necessarily have about what the others will do. Jokes about game theory's recommending that finance ministers toss coins to decide precisely when to devaluate are therefore misplaced: finance ministers can achieve exactly the same effect precisely as they always have, i.e. by using a committee of economic and financial experts.

The idea that mixed strategies can be "purified," i.e. that nobody need consciously be randomizing in order for mixed strategies to be meaningful, has a long history. The first expression of this idea in an *eductive* context seems to be due to Harsanyi (1973). A more recent reference is Aumann, Katznelson, Radner, and Rosenthal (1981). What is offered in this section is therefore only a new twist on an old story. Of course, in an *evolutive* context, all is much easier since a mixed strategy can then be seen simply as a summary of the behaviors current in the population as a whole.

6.5 Learning

Sections 6.3 and 6.4 were confined to the discussion of *static* contests. In this section contests with some *dynamic* structure are studied. As soon as time enters the picture, the problem of learning has to be faced. The traditional approach is to use Bayesian updating, often within the conceptual framework created by Harsanyi (1967–8) for his theory of "games of incomplete information." This section also takes these tools of analysis as basic.[42] Where it differs from the traditional approach is in its enlargment of the range of phenomena about which the players may learn as the game unfolds. In particular, the players may learn about *their opponents' thinking processes* from observing their behavior.[43] This includes not only their opponents' beliefs but also the manner in which these beliefs are constructed. A major complication is that such a view requires abandoning the principle (whether explicit or implicit) that deviations from equilibrium behavior should be taken to be *uncorrelated*. The examples of part I were intended largely for the purpose of demonstrating that, although the principle may be defensible in some evolutive contexts, it is *not* intuitively satisfying in an eductive context.

The issues considered here are obviously very relevant to "signalling games," i.e. games in which the actions chosen by the players may serve to signal important information to the opponents, usually in respect of all the signaller's own future intentions in the game.[44] Some brief discussion of these issues appears in section 6.6. At this point, I want only to stress that the approach advocated here *rejects* an assumption which is explicit or implicit in much of the literature: namely, that beliefs can be seen as the object of strategic *choice*. This seems to me to make nonsense of the fundamental Bayesian insight that preferences and beliefs can be *separated*. Indeed, one might go so far as to say that the purpose of studying rationality is *so that* beliefs can be formed *without* their being distorted by preferences. The approach advocated here derives from a different line of research, namely that developed by those who have worked on the value of *reputations* in a game-theoretic context (e.g. Kreps and Wilson, 1982b).

In this work, it is given that the opponent may not be "rational." Instead it is assumed that a small initial probability exists that the opponent is a type of player who is "irrational" in a highly specific manner. This probability is revised as observations of the opponent's behavior become available. Matters are complicated by the fact that it will often be worthwhile for a "rational" opponent to consider aping the behavior of an "irrational" opponent in the hope of being wrongly identified. Interest centers around the precise extent to which such "hustling" is worthwhile and its role in generating behavior which is markedly different from that which occurs when the possibility of "irrationality" is ruled out altogether.

The twist on this story offered here is that, once the idea of perfect rationality has been put aside, the same conceptual framework can be employed *without* postulating the possible existence of players who are downright irrational. Instead, one may begin with an array of *types* of player, with each type representing a different specification of the factors not tied down by the assertion that a player is "rational." This allows a wider range of behavior to be reconciled with "rationality." albeit only "approximate rationality," then would otherwise be possible. Notice, however, that it is *not* claimed that *all* behavior is capable of being reconciled with "rationality." My own opinion is that a really good theory would include, not only a clear picture of the types of machine to be regarded as approximately rational, but also a clear view on the nature of the *defects* likely to afflict such machines so that a "minimal deviation" hypothesis is *always* available to "explain" off-the-equilibrium-path actions. If an evolutionary origin for the "rational" machines is envisaged, such defects can be identified with unimproving mutations. What needs to be emphasized is that, whether or not the opponent is thought to be "rational," deviations from predicted play in this framework will *not*, in general, be treated as conveying *no* information about possible future behavior.[45]

For simplicity no further account will be taken of the possible existence of "irrational" types of player, although this imposes a severe restriction on the class of games to which the discussion is applicable. This allows attention to be focused on the source of the variation in the different types of rational player. Recall the argument of section 6.3 which asserts that a "rational" machine will necessarily incorporate a stopping rule and a guessing algorithm. Both provide possible sources of variation among types but it is the guessing rule which seems important here.[46] Also remember the distinction made in section 6.3 between subjective and objective data. The procedure proposed here is to continue to treat subjective data, i.e. data a machine can obtain by examining its own operation, as common among the players. In practice, this means that, in simulating another machine, a machine works on the hypothesis that the other machine has the same program as itself.[47] However, in this section, the objective data will not be treated as part of this program but as a private input to the program. The important point is that this objective input may vary between machines and that it is to this variation that differences in the basic guessing rule are to be chiefly attributed.

The illustrative example employed in sections 6.3 and 6.4 (see figures 6.1, 6.2(a) and 6.3(a)) treats these differences in the basic guessing rule as being sufficiently small as to be negligible. This seems reasonable as a hypothesis if it is supposed that the players have a very large amount of relevant objective data, since one can then appeal to some variant of the law of large numbers. However, in section 6.3, it was argued that the interesting case for an eductive analysis is that in which the amount of relevant objective data is small. In this case the differences between different machines must be expected to be significant and the machines' programs must be expected to recognize this fact.

Figure 6.2(b) is an attempt to adapt the illustrative example used previously to take proper account of this more general situation. The machines are assumed to recognize n different possible types of machine, distinguished by the differing objective data they may have received. The objective data received by a machine of type i is assumed to determine a set of beliefs for that machine about the objective data (i.e. the type) of the opponent.[48] The machine then employs these beliefs to generate a basic guessing rule G_0 to be used in predicting its opponent's behavior. In figure 6.2(b) p_{ji} is the probability attributed by a type i machine to the event that the opponent is type j. These probabilities are taken to be part of the subjective data of a machine since they are obtained as a by-product of the manner in which a machine *processes* its objective data.[49] The notation $G_1^j(y)$ represents an optimal response by a machine of type j occupying the role of player y given that such a machine predicts the behavior of its opponent x using the guessing rule $G_0^j(x)$. Since a machine of type i does not know the type of its opponent, it assesses these optimal responses with the weighted average

$$\tilde{G}_1^i(y) = p_{1i}G_1^1(y) + p_{2i}G_1^2(y) + \ldots + p_{ni}G_1^n(y)$$

Assuming, as in section 6.3, that G_0^i is already a good basic guessing rule for a machine of type i, so that the probability r_0 may be taken to be small, a self-correction by a machine of type i would be to replace G_0^i by

$$H_0^i = (1 - r_0)G_0^i + r_0\tilde{G}_1^i$$

Figure 6.3(b) shows how this modification is reflected in the dynamics for the simple zero-sum game studied in section 6.4. Only two types are considered and the points labeled 1 and 2 represent initial basic guessing rules from which the bootstrapping process begins. Each type attaches probability 3/4 to the event that the opponent is of the same type as itself. As in the case of only one type described in section 6.4, both trajectories converge on the unique Nash equilibrium. Thus, although the two different types come to the game with different experiences, introspection leads them to approximately the same final basic guessing rule.

Recall that an eductive libration takes place entirely *inside* a single player's head in a static game. The example is therefore describing the manner in which introspection can lead rational players to reliable conclusions about rational opponents even though the premises from which the opponent began may be uncertain. It is some such intuition that lies behind the classical defense of Nash equilibrium. However, this classical defense requires that the predictions each player makes about his or her opponent's play are common knowledge.

Figures 6.3(c) and 6.3(d) indicate situations with two and four types respectively. In figure 2.3(c) each type, rather eccentrically, attaches probability 1/10 to the event that the other player is the same type as itself. Nevertheless, convergence to the unique Nash equilibrium is still attained. To obtain an example without convergence to the Nash equilibrium, it is necessary to consider even more bizarre initial conditions, as in figure 6.3(d). Here type 4 is certain that her opponent is type 3, type 3 that his opponent is type 2, type 2 that her opponent is type 1, and type 1 that his opponent is type 4.

So far the approach of section 6.3 for static games has been generalized to take account of different types of "rational" player. Obviously, what has been said has relevance to the problem of selecting equilibria in static games when *many* rival equilibria exist. Harsanyi and Selten (1980, 1982) make this problem the chief motivation for studying their closely related tracing procedure. However, I am reluctant to follow Harsanyi and Selten in making a sequence of heroic assumptions from which to proceed[50] and hence this issue will be put aside in favor of a study of the implications of the approach for dynamic games. This study is taken up shortly. But, before leaving static games altogether, it will be instructive to look briefly at the following version of the "battle of the sexes" in order to comment on the relevance in this context of Aumann's (1974, 1987) notation of a *correlated equilibrium.* For a discussion of the traditional viewpoint in this connection, see, for example, chapter 4.

The game matrix for the "battle of the sexes" is shown in figure 6.6. The game has three Nash equilibria. In figure 6.4(a), which is analogous to

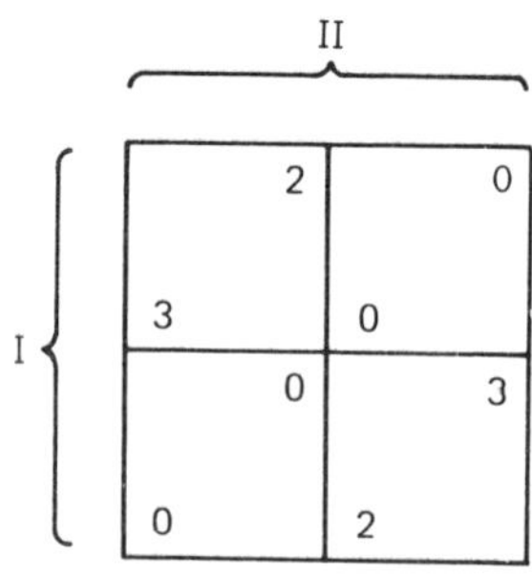

Figure 6.6

figure 6.3(a), these are located at the points (0, 0), (1, 1), and (2/5, 3/5). The first of these, for example, corresponds to each player's using his or her first pure strategy. If only one type of player is admitted (i.e. both players begin with the *same* objective data), then introspection will necessarily lead both players to nearly the same final basic guessing rule, and this will be one of the three Nash equilibria. In figure 6.4(a) the Nash equilibrium to which they are led is (0, 0). If their objective data had led them to place their initial basic guessing rule at any point (p, q) with $p + q < 1$, the result would have been the same. With $p + q > 1$, the resulting Nash equilibrium would have been (1, 1). Figure 6.4(b) shows that the Nash equilibrium to which the players are led when the initial basic guessing rule satisfies $p + q = 1$ is (2/5, 3/5). Harsanyi and Selten's "tracing procedure" selects the last of these as the "solution of the game." Their defense of this result depends on locating their equivalent of an initial basic guessing rule suitably. One might, for example, argue that, since the game exhibits a certain symmetry, then the initial basic guessing rule should exhibit the same type of symmetry. This would locate it on the line $p + q = 1$ and hence one would be led to select the same Nash equilibrium as the Harsanyi–Selten theory. However, I find such an argument too weak to be acceptable. This is partly because I see no particular reason for insisting on a priori symmetry requirements in the players' objective data and partly because, even for situations in which such symmetries might be defensible, the slightest perturbation of (p, q) from the line $p + q = 1$ destroys the conclusion entirely.

It is, indeed, far from clear that attention should be confined only to Nash equilibria in this context. The objective data that types receive will, in general, be *correlated*, since it is obtained from observing the behavior of the *same* population of players. It is therefore not surprising that, if the process of refining guessing rules converges, then it does not necessarily converge to a Nash equilibrium of the game although it always converges to a *correlated equilibrium* of the game. Correlated equilibria are relevant whenever both players are partially informed about the outcome of some exogenous random event and condition their choice of strategy on this information. The requirement for equilibrium is then the same as that proposed by Nash: no player must have an incentive to deviate from his or her choice, provided that the other player does not deviate. A Nash equilibrium is a correlated equilibrium for which each player's information about the exogenous event is *independent* of the other player's information. Otherwise a correlated equilibrium may differ from a Nash equilibrium.[51]

As an example, consider the case when the exogenous random event consists of whatever prediction is written in a game theory book about how the "battle of the sexes" will be played by rational individuals in the same circumstances as the two players currently about to play the game. The event is random in the sense that neither player is assumed to have read the book and so what it actually says is unknown to them. But each player is told what the book predicts that he or she will do by friends who have read the book. However, the friends neglect to provide information on what the book has predicted that the opposing player will do. Suppose now that, *before* receiving information from a well-read friend, it is common knowledge that each player attaches probability 3/8 to the event that the book predicts the northwest cell of the "battle of the sexes" payoff matrix as the outcome of the game, 3/8 that it predicts the southeast cell, and 1/8 for each of the other cells. It is now a correlated equilibrium if, *after* communicating with their well-read friends, each player simply chooses the strategy his or her friend reports as being the prediction the book makes for that player. For example, if the game theory book predicts for player I that he will choose his first pure strategy, then player I will use this prediction to update his prior probabilities for the predictions that the book may have made for player II. The revised probabilities will be 3/4 for her first pure strategy and 1/4 for her second pure strategy. Thus, if player I chooses his first pure strategy, he expects a payoff of $(3 \times 3)/4 + (0 \times 1)/4 = 9/4$. If he chooses his second pure strategy, he expects a payoff of $(0 \times 3)/4 + (2 \times 1)/4 = 2/4$. It is therefore optimal for him to follow the prediction made about him by the game theory book, provided that he believes the other player will do the same.

The final guessing rules in figure 6.4(c) correspond to the correlated equilibrium described above. In this example, each of the two types attaches a probability of 3/4 to the event that the other player is of the same type as itself. But, unlike the situation in figure 6.3(b), the final guessing rules of the two types are *not* the same, nor is their behavior in the game. Type 1s always use their first pure strategy whether they are player I or player II. Types 2s always use their second pure strategy. Note that, although the result of the game is not a Nash equilibrium, everybody is always optimizing *according to his or her beliefs*.

Figure 6.4(d) is a final comment about symmetry requirements on the initializing objective data. Each of the four types attaches probability 1/4 to the event that the opponent is any one of the four possible types. The initial basic guessing rules are also symmetric *except for very small displacements*. The result, however, is far from symmetric. Each type ends with the *same* guessing rule, corresponding to the Nash equilibrium at (0, 0).

To summarize the conclusions to be drawn from the preceding brief study of the "battle of the sexes," the theory offered here envisages the problem of equilibrium selection as being determined by the objective data with which the players come equipped to the game. Blanket assumptions, and, in particular, sharp symmetry assumptions concerning the nature of this objective data, seem ill advised. Moreover, not only Nash equilibria, but also correlated equilibria, may be generated, thus providing some support for Aumann's (1974, 1987) view that correlated equilibria have an important role in this context. These conclusions would certainly not be very promising if one were hoping for a method of analyzing games which was independent of the environment in which the game is played. But this aim seems to be both hopeless and misguided.

Thus far in this section, attention has been confined to static games so that the ideas of the preceding sections can be extended to the case in which players may be of many different types. In dynamic games, this extension of the theory has simple but important consequences. In a dynamic setting, the beliefs attributed to the various types of machine about the type of their opponent are *vulnerable to Bayesian updating*, as data are supplied by the opponent's choice of actions in the game. Recall that, in the illustrative example of figure 6.2(b), the beliefs of a machine of type i are encapsulated in the probabilities $p_{1i}, p_{2i}, \ldots, p_{ni}$. However, such an example will no longer suffice for a dynamic game because the fact that Bayesian updating takes place is not modeled. In a dynamic game, not only will beliefs be updated over time, but this updating will be simulated by those seeking to make predictions about what might happen in the *future*. This, in turn, will force attempts to be made to simulate the thinking processes which led to decisions made in the *past* so that deductions can be made about what predictions about the future those decisions were based on. But this will sometimes provide an incentive to choose actions precisely because they render such deductions difficult to make accurately. No attempt will be made to unravel the complexities involved, except for the very simple examples given in the remaining sections. Instead, this section will be closed with a list of the operations it is envisaged that an appropriate machine would need to carry out on arrival at an information set.

Imagine that a machine receives a message signifying that information set h has been reached. At this time, it will be equipped with a probability distribution λ_h over the set of possible types of its opponent. It then proceeds as follows:[52]

Operation 1	For each possible type, obtain predictions of the behavior of that type at all relevant information sets (past, present, and future).
Operation 2	Update the previously held probability distribution λ_h using Bayes's rule, conditioning on the event that h has been reached.
Operation 3	Use the updated value μ_h of λ_h to assign probabilities to each node in h and to each action available at an information set which follows a node in h.
Operation 4	Use the results of operation 3 to assign Von Neumann and Morgenstern utilities to each action available in h.
Operation 5	Choose an optimal action.

Two things need to be emphasized. The first is that operation 1 involves a very heavy task. To simulate another type at another information set requires simulating the type's use of operations 1–5. But as soon as the simulation of operation 1 is attempted, simulations of simulations become necessary; and so on. Section 6.7 looks at some of the implications in a very simple game. The other point requiring emphasis is that the set of types considered must be large enough to ensure that all information sets are reached with positive probability. This will not usually be possible if only "rational" types are considered.

6.6 Beer versus quiche

In this section the ideas of the previous section are applied, *in an informal way*, in two simple signaling games. It is *not* claimed that the insights offered are original. All that is proposed is to express the insights within a more coherent framework than is usual.[53]

The first game is Kohlberg's dalek example, which was used in part I as a hook from which to hang some general discussion of game-theoretic ideas. The rules of the game are embodied in figure 6.7(a). Player I begins at node x and has a choice of action A or action D. If he chooses A, the game ends and each player receives a payoff of one unit. If he chooses D, then player II has to choose between actions l and r. Whatever her choice, player I

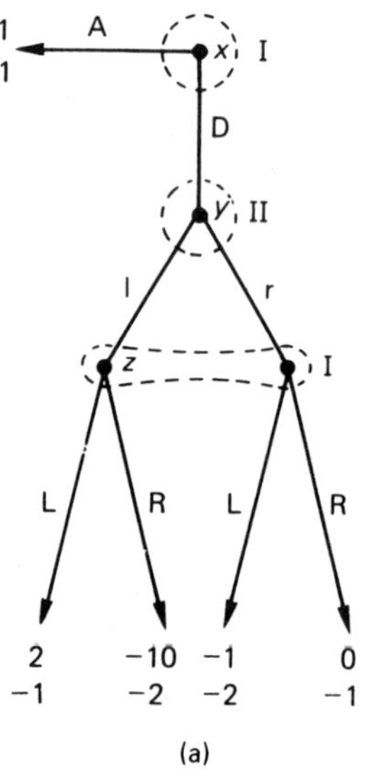

(a)

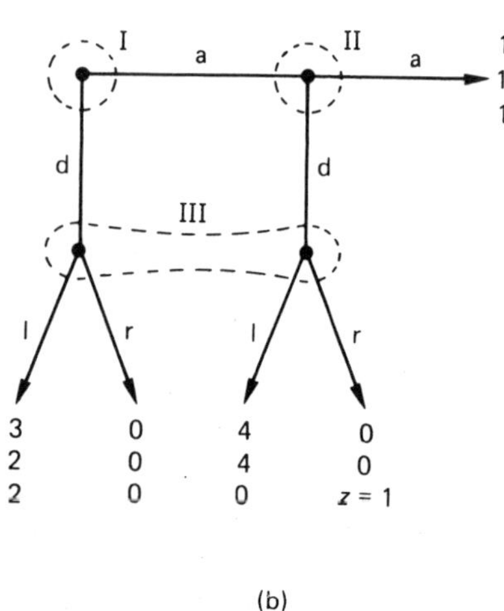

(b)

Figure 6.7

then has to decide between actions L and R. The two nodes at which this decision may be exercised are enclosed in an "informationa set," labeled z, to indicate that player I does not know, in choosing between L and R, what choice player II made between l and r. It may help to think of the game as one in which player I may choose between a sure payoff of one unit and playing a game of the "battle of the sexes" variety with player II.

The insight to be studied is that the play of D by player 1 at information set x will serve as a signal to player II that he intends to play L at information set z in the expectation that player II will play l at y. Such a "forward induction" argument would serve to establish the sequential equilibrium ($\langle$D,L$\rangle$,l) at the expense of its rival ($\langle$A,R$\rangle$,r). (If clarification is needed, see part I, section 5.2.)

The appropriate intuition may be defended directly, without resort to mathematical legerdemain, as follows. An optimizing player I can only choose D if his guess at player II's behavior assigns a probability of 2/3 or more to her choosing l if node y is reached. But, if he is of a type with such a guessing rule, then he will play L at information set z, since he will receive no supplementary information in passing from x to z. In seeking to simulate player I's attempt to predict her behavior should she get to play, player II will be guided by her objective data. This will lead her to a probability distribution over possible types of player I. Each of these types will have his own guessing rule (and beliefs about the type of player II). Whatever player II's original probability distribution may be, it cannot but be updated[54] after the observation of the play of D to a distribution which assigns zero probability to those types which, acting as player I, would choose A. The other types all choose L at z and hence player II optimizes by choosing l at y. Since all types of player I will simulate this behavior of player II, all types of player I will therefore play D. This generates the equilibrium ($\langle$D, L$\rangle$,l).

The issue of whether some types of player I may be categorized as "hustlers"[55] is irrelevant to this conclusion. If hustling types exist, their simulation of player II's response to their attempt at "deceit" must have indicated that the deceit would be successful.

There are at least two weaknesses in this argument. The first is that it assumes that the appropriate eductive librations will converge. The second, which is more serious for applications, is that irrational players have been excluded. But in real life, one might well have good reason to suppose that the event that the opponent is rational but equipped with an unlikely data set would be much less probable than the event that he is just plain stupid.

The second signaling game is the "beer versus quiche" example of Kreps (1985). The version studied here is that given in figure 6.8(a). The payoffs offer a bonus of +1 to the soft player 1 for eating quiche and a bonus of +1 to the tough player II for drinking beer, in accordance with the adage,

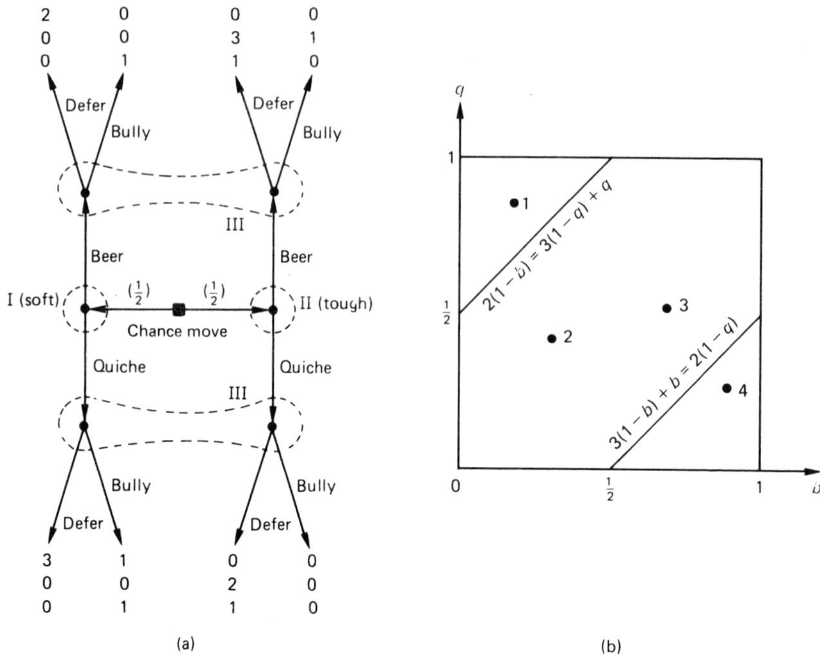

Figure 6.8

"Real men don't each quiche." Two rival classes of equilibria need to be examined.

The pure strategy representative from the first class requires player III to treat quiche as signaling that he is playing the soft guy and beer as signaling the tough guy. Player III therefore bullies after observing quiche and defers after observing beer. Both the soft and the tough guys therefore optimize by signaling beer. The pure strategy representative from the second class requires player III to treat quiche as signaling that he is playing the tough guy and beer as signaling the soft guy. Player III therefore bullies after observing beer and defers after observing quiche. Both the soft and tough guys therefore optimize by signaling quiche.

The equilibria in the second class seem perverse. One argument that could be used in an attempt to exclude them is based on Banks and Sobel's (1985) notion of a "divine equilibrium."

Figure 6.8(b) indicates possible guessing rules for player III's behavior. A point (b, q) is to be identified with a rule which assigns probability b to the event that III will bully given that beer has been signaled, and q to the event that III will bully given that quiche has been signaled. For simplicity, it is assumed that only four types of player occur with positive probability and that these generate the guessing rules labeled 1, 2, 3, and 4 in figure 6.8(b). It is optimal for the soft player I to signal beer if and only if he is type 1. It is optimal for the tough player II to signal beer if and only if she is type 1, type 2, or type 3.

Suppose that player III observes the signal "beer." How does this affect the probabilities he attaches to the event that the signaler is the soft player I or the tough player II? Bayes's rule,

$$\frac{\text{prob(soft}\,|\,\text{beer)}}{\text{prob(tough}\,|\,\text{beer)}} = \frac{\text{prob(beer}\,|\,\text{soft)prob(soft)}}{\text{prob(beer}\,|\,\text{tough)prob(tough)}}$$

$$= \frac{\text{prob(type 1)}}{\text{prob(type 1 or type 2 or type 3)}} < 1$$

provided that the type of a player is independent of his or her role in the game.

It follows that it is optimal for player III to defer if beer is signaled. A similar argument shows that bullying is optimal if quiche is signaled. Knowing this, both the soft and the tough player signal "beer."

The argument serves to illustrate two points. The first is the value of admitting different types of player. The second is the danger of discussing equilibria in a vacuum without reference to the libration process by means of which equilibrium is achieved. If the argument is valid, then the only correct prediction for player III's behavior is $(b, q) = (0, 1)$. Thus, if types 2, 3, and 4 occur with positive probability, then these types are not very rational in that their predictions are badly mistaken (although, since they were assumed to have optimized on the basis of their beliefs, they certainly are "Bayesian rational"). It may be that the environment in which the game is played is such that types of this kind can never be excluded. If so, then all is well with the argument. But for an analysis in which such "irrational types" are not to be admitted, matters are not so clear. It is evident that a libration of the kind discussed in section 6.4 will sweep primitive guessing rules located at points 1, 2, 3, and 4 towards the northwest corner of figure 6.8(b) and that this process must continue until the central region is swept clean. This will "typically" leave all guessing rules in the northwest region of figure 6.8(b) *but not always*. Thus, unless one is willing to be more specific about the nature of the objective data on which the players base their calculations, there seem to be no good reasons for ruling out the "perverse equilibria" altogether in the absence of "irrational types." All that is genuinely available is a strong presumption that the perverse equilibria will not be observed unless the preliminary objective data contain a heavy bias in their favor. But, for most purposes, this is perhaps enough.

6.7 The horse example

Selten's (1975) horse game is reproduced from part I as figure 6.7(b). It is orthodox to reject the Nash equilibrium (d, a, l) in this game in favor of the Nash equilibrium (a, a, r) on the grounds that the former is neither a perfect nor a sequential equilibrium. In part I, section 5.4, the validity of this orthodoxy was challenged for the eductive context but not for the evolutive context. The essence of the argument was that deviations from equilibrium should not be treated as being automatically uncorrelated with other deviations. This point is easier to swallow for Kohlberg's dalek example of the previous section than for the horse example, but the logic of the situation applies here also.

The argument given in part I, section 5.4, has the character of a *reductio ad absurdum* in that its aim was to demonstrate inconsistencies in the traditional argument rather than to make a genuine claim for (d, a, l). That is, the challenge was to the orthodox *argument* rather than to the orthodox *conclusion*. In this section, some attention is given to the substantive question. The discussion is closely related to the idea of a correlated equilibrium as introduced by Aumann (1974) and developed by Myerson (1984) and others (see section 6.5).

The aim is to tell a simple story about the environment within which the horse game might be played that can lead to the use of equilibria with orthodox theory rejects. I am unable to find an adequately simple story with the payoffs given by Selten.[56] The first step is therefore to change the same so that the payoff vector following the history $\langle a, d, r\rangle$ ceases to be (0, 0, 1) and becomes (0, 0, z). The Nash equilibria in the new game remain the same as in the old game. To be precise, they fall into two classes. The first class has player I and player II choosing a and player III choosing r with probability $p \geqslant 3/4$. The second class has player I choosing *d*, player III choosing l and player II choosing d with probability $q \leqslant 2/3$. For all values of $z > 0$, a traditional analysis accepts all equilibria in the first class but rejects all equilibria in the second class on the grounds that player II's plan of action is then "irrational" because *if* she found herself called upon to play she would not choose a and hence a payoff of 1 because the choice of d gives her a payoff of 4, *assuming* that player III chooses l as the equilibrium requires (see part I, section 5.4). The point of the story which follows is that such a rejection cannot be properly justified if z is sufficiently large. The actual value of z taken is 5.

Before the game is played, each player receives one of two possible pieces of objective data, labeled 1 and 2. This determines the type of a player. To make the conclusions comparable with the results of a traditional analysis, it is assumed that the receipt of a piece of objective data by a player will induce the player to believe that it is highly probable, although not altogether certain, that the other players will have received the *same* objective data. A libration carried out in such circumstances will "typically" lead players to a Nash equilibrium, but it will *not* necessarily lead type 1 and type 2 players to the *same* equilibrium. Instead, it will simply be assumed that type 1 players are led to a particular equilibrium in the first of the two classes of equilibria, and that type 2 players are led to a particular equilibrium in the second class. It will be argued that, within this framework, the proposed plan of action of a type 2 person in the role of player II *cannot* be dismissed as "irrational."

The probability p with which player III chooses r in the type 1 equilibrium and the probability q with which player II chooses d in the type 2 equilibrium will depend on the prior beliefs that individuals have about the manner in which the objective data is generated (before they receive any such data). Figure 6.9 indicates the specific prior beliefs which will be adopted. In this figure, $\delta = 1/5$, $\bar{\delta} = 1 - \delta = 4/5$ and $\varepsilon > 0$ is very small indeed. (With this choice for δ, it will emerge that p needs to be 15/16 and q needs to be approximately 1/2.)

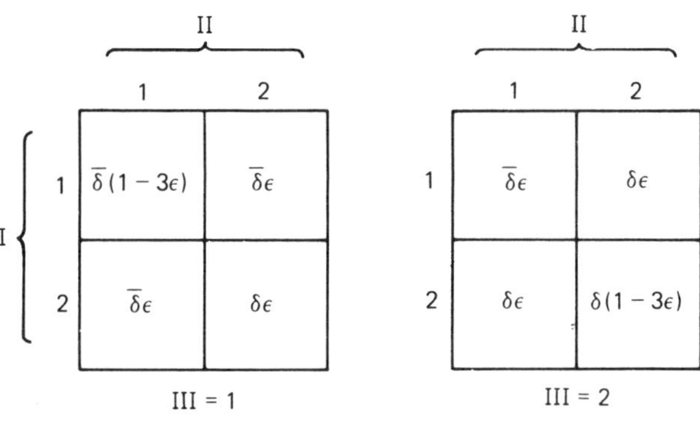

Figure 6.9

The example does not require that an explanation of these beliefs be offered. However, so long as the story is not taken too literally, it may be helpful to think of a conference of game theorists who decide on whether a type 1 or a type 2 equilibrium is to be recommended. The players do not directly observe the outcome of the conference. Instead, they receive a letter informing them of the outcome. But the administering computer is vulnerable to "trembling-hand" errors and, each time that a letter is sent, there is a probability ε that the wrong letter is sent. The given matrix is then obtained by assuming that the players will attach a prior probability of $\delta = 1/5$ to the event that the conference will select the type 2 equilibrium and then suppressing quadratic and higher terms in ε.

The matrix of prior beliefs is used to calculate the following probabilities:

$$P_1 = \text{prob}(\text{I} = 2 | \text{III} = 1) \simeq \varepsilon/\bar{\delta} = 5\varepsilon/4$$

$$P_2 = \text{prob}(\text{I} = 1 \text{ and II} = 2 | \text{III} = 1) \simeq \varepsilon$$

$$P_3 = \text{prob}(\text{III} = 1 | \text{I} = 1 \text{ and II} = 2) = \bar{\delta} = 4/5$$

With a suitable value of p in the type 1 equilibrium and of q in the type 2 equilibrium, it will be shown that *all* planned behavior in this set-up is entirely rational in the traditional sense. No difficulty arises in this respect about the planned behavior of persons of type 1 in roles I and II, nor about the planned behavior of persons of type 2 in roles I and III (provided that $p > 3/4$ and $q < 2/3$). They commence the game believing that it is nearly certain that the other players have computed the same equilibrium as they have themselves, *and they receive no countervailing evidence.* Hence the standard Nash equilibrium argument applies to their strategy choices. As always, the difficulty is with what happens at an information set which "cannot be reached in equilibrium." In the current context, this means that attention needs to be directed at a type 2 person in the role of player II and at a type 1 person in the role of player III.

Taking the former first, what does a type 2 person in the role of player II deduce from the event that she is called upon to play? Her explanation, in the current framework, must be that player I is of type 1 (and not of type 2 as she thought highly probable to begin with). This information is relevant to what she now ought to believe about player III. Her posterior belief that player III is of type 1 now becomes $P_3 = 4/5$. Since type 1s in the role of player III choose r with probability p and type 2s in the role of player III always choose l, her expected payoff from choosing d is therefore

$$P_3(1-p)4 + (1-P_3)4$$

To justify a mixed strategy choice for a type 2 person in the role of player II, it is therefore necessary that this expected payoff be equal to the payoff of 1 she obtains by choosing a. The requirement on p is that $p = 15/16$.

Next consider what a type 1 person in the role of player III deduces from the event that he is called upon to play. He can explain this event in two possibly ways. Either player I is of type 2, or player I is of type 1 and player II is a type 2 person whose randomization[57] has led to the choice of *d*. The first explanation places player III at the left-hand node in his information set; the second explanation places him at the right-hand node. The respective conditional probabilities for these two events are $P_1/(P_1 + qP_2)$ and $qP_2/(P_1 + qP_2)$. To justify a mixed strategy choice for a type 1 person in the role of player III, it is necessary that the expected payoff from the choice of l be equal to that from the choice of r. It is here that the modification of payoffs in Selten's original example is required. With $z = 5$, the requirement is that

$$2P_1 = 5qP_2$$

and hence q must be approximately 1/2.

In summary, the analysis envisages that two different types of player will make two different analyses of the game. Both types think it nearly certain that their analysis is "correct." But if they observe a deviation, they reassess this opinion. In particular, if a type 2 person in the role of player II observes a deviation from what she thought to be "correct" play on the part of player I, she does not dismiss this as irrelevant to her prediction of what player III will do, as required by a conventional analysis. Nor does the assume, as in part I, section 5.4, that it *must* be she who is the "odd man out." Instead, she carries out a standard Bayesian updating of her beliefs about the type of player III. With $z = 5$, she then has no grounds for altering her original plan to play d with probability q.

The argument is vulnerable to criticism on various grounds. In particular, the manner in which the piece of objective data received by a type is translated into a recipe for action is entirely unmodeled. Moreover, a more realistic framework would envisage many possible pieces of objective data, so that the probabilities attached to alternative recipes could be treated as endogenous to the model. However, perhaps the model is not so primitive as to obscure the basic point altogether.

Finally, something needs to be said about an *evolutive* analysis of the game. This is connected to what was said in section 6.5 about the distinction between those situations in which there is little objective data available and those in which there is much. It is the former situation which was said to be of interest for an eductive analysis. When there is a large amount of objective data available, the probability that two different types will have significantly different guessing rules and belief systems must be expected to be very small.[58] But once this probability has become small enough, it will cease to make sense to treat the possibility that a deviation has an *irrational* origin as negligible. A different analysis will then be necessary; and this will need to be an analysis based essentially on evolutive principles.

6.8 Conclusion

This last section is an attempt to draw together some of the more positive strands that have been proposed in the preceding sections. Figure 6.10 illustrates the general scheme suggested for a strictly eductive analysis when no account need be taken of the possible existence of "irrational" types of player. It should be emphasized that this is *not* a new "equilibrium concept." All that is offered is a skeleton that would need to be fleshed out with

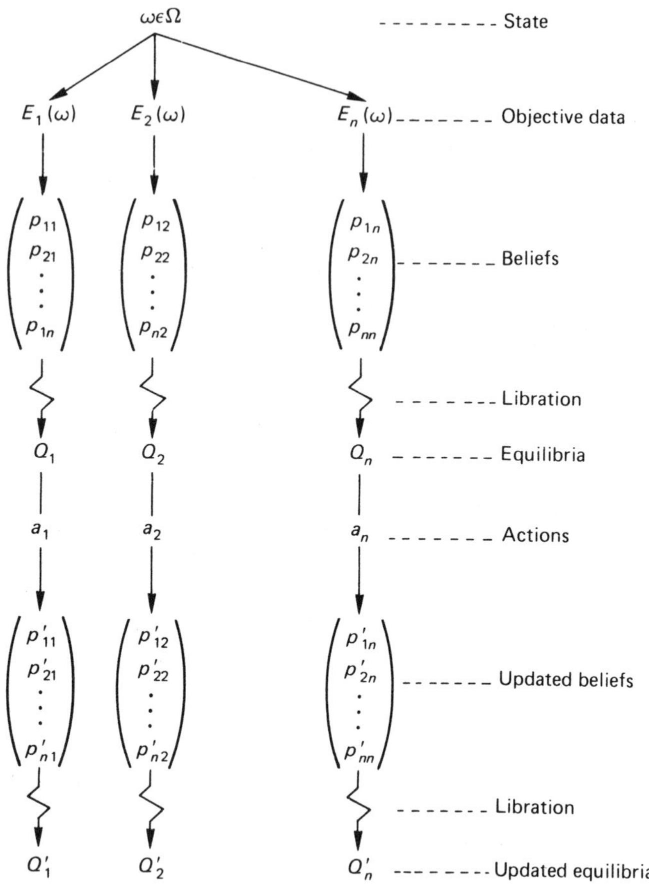

Figure 6.10

numerous modeling judgments before it could serve in such a role. A major drawback is the fact that the exclusion of "irrational types" severely limits the potential range of application of the scheme.

The various stages in figure 6.10 are itemized below.

1 *State*: The symbol Ω represents a set of possible states of the world. Nature chooses a particular state $\omega \in \Omega$. It is assumed that a common prior on Ω is common knowledge.

2 *Objective data*: The players are drawn at random from a population of different types of person. The types are distinguished by the fact that, although they *process* data in essentially the same way, they begin with different, but possibly correlated, pieces of information. In state ω, type

i observes event $E_i(\omega)$, which is identified with the objective data described in sections 6.3 and 6.5. The nature of the events $E_i(\omega)$ is taken to be common knowledge.

3 *Beliefs*: Each type of person forms beliefs about the type of a person drawn at random from the population to fill the role of another player. In particular, p_{ji} denotes the probability attached by a type i person to the event that another player is of type j. The numbers p_{ji} are taken to be common knowledge.

4 *Equilibria*: Each type of person reasons his or her way to an equilibrium before there is any action in the game at all. The equilibrium will not be anything simple except in trivial examples (like those studied in sections 6.6 and 6.7). A complicated backwards induction will be involved which begins by calculating equilibria after penultimate actions in the game for all possible beliefs that types might then have about the types of other players. Although backwards induction of a kind is used, the result will therefore not look anything like a subgame-perfect or trembling-hand perfect equilibrium.

5 *Action*: Now comes the first action in the game. If the player who moves is of type i, he or she is assumed to take action a_i is required by the calculated equilibrium Q_i. The calculation of this equilibrium will, of course, take into account the effect that the action a_i is expected to have on the beliefs of other types about the types of their opponents.

6 *Updated beliefs*: The scheme of figure 6.10 is adequate only for two-player games in this respect, although the appropriate generalization will be evident. If the player who did not take the first action obtains some information about the action taken by the other player, then his or her beliefs must be updated.

7 *Updated equilibria*: New equilibria must now be calculated based on the updated beliefs.

8 And so on.

In spite of the drubbing delivered to naive Bayesianism in part I, section 5.6, it will be observed that, in the end, Bayesian ideas have not disappeared from this scheme. They still provide the essential framework for the analysis. And this includes what Aumann calls the "Harsanyi doctrine": namely, there is common knowledge of common priors, so that differences in behavior between two individuals faced with identical problems[59] are attributable principally to differences in their past experiences rather than to fundamental differences in the manner in which they solve problems.[60] The vital point at which naive Bayesianism is replaced by something a little more sophisticated (in that explicit account is taken of the "massaging process" described in part I, section 5.6) is the manner in which equilibrium is envisaged as being achieved (the libration interludes in figure 6.10). Otherwise, the ingredients of the scheme will be familiar to game theorists, although this will not be true of the manner in which they are mixed and stirred. A "trembling hand" will typically be involved in the choice of the state ω. The correlation between the observed events $E_i(\omega)$ is borrowed from Aumann's notion of a "correlated equilibrium." The use of types and, in particular, the treatment of beliefs is derived from Harsanyi's theory of incomplete information. "Backwards induction" appears in the calculation of equilibria. Some species of "forwards induction," as envisaged by Kohlberg and Mertens, is involved in the updating of beliefs and the recalculation of equilibria. But, even though the scheme is proposed only for the strictly eductive analysis of a restricted class of games (those for which all behavior can be explained "rationality"), it is still palpably inadequate in spite of the panoply of game-theoretic concepts which it incorporates.

Finally, a moral should be appended. Much of this two-part study has been overtly rhetorical. Even those passages which are not overtly rhetorical are covertly so, since the basic aim has been to demonstrate that, no matter how insistent an analyst may be that his ivory tower is erected exclusively to the glory of strictly normative gods, *nevertheless* procedural questions *cannot* be evaded without generating confusion and inconsistency. Costly, or bounded, rationality has to be a feature of *any* worthwhile theory, whether normative or positive. But costly rationality is not a subject that can be tackled successfully by abstract mathematical reasoning alone. The nature of the equilibrium reached will, in general, depend on factors which cannot be determined from an armchair. In particular, the previous *history* of the system must be expected to matter and thus any serious investigator must have at least some knowledge of the relevant *empirical* evidence.

NOTES

1 Of sufficient complexity (Gödel's theorem).
2 The theory of the organization of the firm is a notable exception.
3 Lewis (1976) offers a more grandiose theory.
4 In what follows later, all sequences of moves are anticipated, in the sense that they are assigned positive probability. What is to be understood is that the observation of a very low probability event will lead to a very substantial revision of the model via Bayesian updating.
5 Playing "badly" in the hope of tempting the opponent into an unwise attempt at exploitation.
6 Although without the implicit assumption that the same results are to be expected in an evolutive setting.
7 It is not denied that common knowledge conditions on beliefs allow a satisfying static analysis of certain static games. But an evasion of the study of the process by means of which beliefs are formed leads to difficulties for dynamic games.

8 The reference is to the Church–Turing hypothesis and includes Turing machines under the general heading of computing machines.

9 Megiddo (1986) summarizes matters elegantly with the oxymoron: "... a fully rational player ... can even decide undecidable problems."

10 Formally, one may think of the failure to make a move by time t as a species of "move" in itself, with appropriate consequences for the game payoffs.

11 Not a meta-game in the sense of Howard (1971), but in the natural sense.

12 Although, in philosophical discussions, the term "bounded rationality" often signals only that procedural questions are not to be ignored entirely.

13 Encouraging because, in spite of allowing complexity to be endogenous and making few assumptions about the cost of calculating, they are still able to generate substantive results.

14 A really good rationality model should be able to solve this choice problem, too – in which case it should pick itself! I have constructed an impossibility theorem for social choice in certain special circumstances using this principle (Binmore, 1975). Perhaps a more general theorem is true.

15 Although the fact that rational agents will treat a bet-laying situation as a game, and hence choose strategically, means that this defense is flawed.

16 Axelrod's (1984) "olympiad" for programs which play the repeated prisoner's dilemma is a good image, aside from the fact that the game menu contains only one item.

17 The physical costs of acquiring and operating hardware seem much less important.

18 When a game is played repeatedly, the term "one-shot" game is used to distinguish the game as played just once from the "repeated game" or "super-game" in which the game and all its repetitions are treated as a whole.

19 Recall that the population of master-program hosts is assumed to be large. Note also that such considerations *will* arise if the game selected for play from the menu of available games happens *itself* to be a repeated game.

20 The discussion which follows leaves the status of this intuition indeterminate. The reason is that there is a tension between the forces which push in this direction and those which maintain the stability of the equilibrium in the face of "mutant invaders." Which wins out would seem to depend on the precise modeling of the dynamics. As always, Harsanyi and Selten (1980, 1982) have considered this point earlier. Perhaps the evolutionary viewpoint of this section will clarify the nature of the tension they see between Pareto dominance and risk dominance when comparing equilibria. It will not, on the other hand, clarify their resolution of this tension which they do not claim to be other than a formalization of their educated intuition of this topic.

21 With some excuse, since it certainly leads to silly conclusions.

22 It is not argued that there will necessarily be a *unique* winner. On the contrary, it is to be expected that a mix of winners will survive who calculate in much the same way but not necessarily with the same precision, nor from precisely the same premises. It may even be that a mix of very disparate winners may survive, held together by an analog of Bayesian equilibrium.

23 In part I, section 5.5, machines are considered which receive, as part of their preliminary input, the Gödel number of an opponent. These were considered

in order to make a philosophical point: namely that the assumption that an opponent's mixed strategy choice can be predicted with precision is questionable. In the current context, it would, of course, be an evasion to postulate that an opponent's Gödel number is known a priori without offering an explanation of *how* it became known.

24 A preliminary version of this chapter (Binmore, 1986) used the term "Bayesian automaton." But it seems that the word "Bayesian" cannot be employed without the risk of being identified with the "naive Bayesianism" criticized in part I, section 5.6. In any case, a change of terminology may serve to signal that the ideas have been developed somewhat. The word "machine" is used in a deliberately wide sense to include any adequately complex computing device.

25 For example, there is the red herring concerning what "rationality" requires in the "pure coordination" bimatrix game in which both matrices are the 2×2 identity matrix. If the strategy labels have no significance for the players, there is nothing to be said since no reason can then exist for supposing it more likely than an opponent will choose one label rather than another. (Note, incidentally, that the *location* of strategy labels in a visual display may well be significant.) If, on the other hand, the labeling of strategies *is* significant, then nothing can be said without a preliminary discussion of how and why the labels are significant.

The example of the preceding paragraph is trivial. At a deeper level, however, there are the "rationality" claims made for such notions as correlated equilibrium (Aumann, 1974, 1987).

26 This is *not* the same as admitting the possibility that mixed strategies may be used. The reference is to uncertainty about which calculation has been employed in determining which strategy is to be used.

27 The answer to this question will clarify why a master-program which writes programs is introduced rather than supposing that all the necessary programs have been written already and stored ready for use. Such a fixed library of programs is then just a poor relation of the "grand book of game theory" to which, following Von Neumann and Morgenstern (1944), game theorists are fond of appealing when pressed for a defense of equilibria in an eductive context. Binmore (1986) evades introducing the idea of a master-program by this method, but only by evoking an "outsider" who intervenes to "improve" the design of machines. This "outsider." or the master-program in the current chapter, substitutes for the *author* of the "grand book of game theory" in the traditional parable.

28 A simple, but costly, method would be to copy the instructions of the master-program to another storage site and then to introduce enough auxiliary instructions to make this usable as a subroutine or procedure. Such a procedure would then correspond to what the biologists call a "simulator" (Monod, 1972).

29 The traditional response is that a "perfect" machine's predictions will *not* fail the criterion. This sweeps under the carpet the question which really matters here: namely, how did the perfect machine get to be perfect?

30 Part I offered Gödel as an authority for such assertions but any philosopher who writes on self-consciousness seems to say something of the sort. An extract to this effect from Schopenhauer is quoted at the beginning of this chapter, but

I prefer Hume's more mundane observation that he often repeatedly decides to rise from his bed but fails to do so, only later to find himself getting dressed with no clear idea of how this came about. For a viewpoint closer to that of this chapter, see Scriven (1965).

31 Observe that to know the program listing of a machine is not to know the results of all calculations of which the program is capable.

32 I see the criteria which Harsanyi and Selten (1982) impose on the prior distribution from which their tracing procedure begins as a brave attempt to systemize their judgments about the extent to which arbitrariness in the choice of their equivalent of a basic guessing rule can be delimited. Much of what they say (although certainly not all) is *obviously* relevant, given such an interpretation.

33 It is tempting to dispense with the stopping rule once its role in establishing the necessity for a basic guessing algorithm is over. Instead, one might go directly to the limit. It is no counterargument that machines cannot "go directly to the limit." According to the Church–Turing thesis, if a pure mathematician has an algorithm for finding exact limits, a machine can be constructed to operate the algorithm. The proper counterargument is quite different. It is simply that working a *finite* number of steps *up* the simulation ladder *from* the basic guessing rule is equivalent to working the same number of steps *down* the latter *to* the basic guessing rule. But working an *infinite* number of steps *up* is equivalent to nothing at all in the original structure attributed to a game-playing machine. This is not to deny that the clearer mathematics usually obtained in the limit may not be relevant, but only to assert that the interpretive problems cannot be ignored, especially since there will normally be many limits to be taken and the *order* in which these are taken will typically be significant.

34 And there is no point in the machine's incurring the cost of monitoring an operation which it is not programmed to correct if necessary.

35 Of course, these probabilities, along with everything else, will be subject to self-correction.

36 Note that such a modeling device does not allow an escape from the difficulties with "perfect rationality" discussed in part I, section 5.5. The machine z cannot then ensure that r is *always* wrong but it can ensure that r is *statistically* wrong. The point here is that, although z cannot calculate r's prediction p when this is generated partly at random, z can calculate the *probabilities* with which the various possible predictions will be made.

37 These are not "nondeterministic automata." Mathematicians have usurped this term for another purpose (Hopcraft and Ullman, 1979).

38 It will be determined by some criterion which renders the machine approximately indifferent to prolonging its self-improvement exercise and to ending it.

39 Especially in the unlikely event that the guessing role happens to make the player indifferent between two pure strategies. For such cases, an *arbitrary* tie-breaking rule was proposed in section 6.3.

40 Although this is not to say that recommending a mixed strategy is *forbidden.*

41 Although this is not to say that randomizing may not frequently be a convenient or cost-effective means for avoiding calculation. Consider, for example, the case of a repeated two-person zero-sum game.

42 The criticisms of Bayesianism in part I, section 5.6, notwithstanding. This

criticism was directed against the inappropriate use of a methodology suitable only for "closed universe" problems when the actual problem is an "open universe" problem. Much of the previous discussion can be seen as an attempt to "close the universe of debate" so as to legitimize Bayesian techniques. Admittedly, however, there remains room for doubt as to the success of this attempt.

43 Indeed, in principle, they might learn things about their *own* thinking process by observing their *own* behavior.

44 As in the "forwards induction" in Kohlberg and Mertens (1986).

45 One might think of the traditional trembling-hand argument as attributing deviations to random electrical fluctuations in the computer *hardware*. The theory envisaged above, on the other hand, would locate the source of deviations in the computer *software*.

46 The importance of the stopping rule is that its existence makes it necessary that there be a guessing algorithm.

47 But recall that some aspects of this program will not be available for self-examination and correction.

48 This is where Harsanyi's (1967–8) incomplete information framework is borrowed. Note the elegance with which problems of "beliefs about beliefs" are treated.

49 This is all that will be offered in defense of the "common knowledge" requirements of Harsanyi's theory.

50 Among other things, I am unhappy about artificial devices introduced to secure convergence. But how are players to evaluate divergent processes?

51 Although Nash equilibrium remains the fundamental concept. A correlated equilibrium is a Nash equilibrium of an *expanded* game for which the first "move" consists of the transmission of the correlated information to the players.

52 As always, the program for carrying out these operations should be seen as vulnerable to self-correction.

53 Although the framework in Kohlberg and Mertens (1986) is certainly mathematically coherent, I would argue that it is far from being conceptually coherent in that the plausibility of their mathematical criteria is hard to evaluate.

54 Assuming that only "rational" players are admitted and that player II concludes that, whatever her own objective data, it remains *possible* that other types may have objective data which lead them to predict that she will play l at y with probability at least 2/3.

55 One might care to label the sending of signal D when equipped with certain categories of objective data as "deceitful." Of course, if the argument given above is correct, such a labeling would be harsh, since, in so far as the signal is that L will be played next, it is accurate. Players of hi-lo five-card stud will recall those situations in which bluffing is well understood to be almost mandatory, and hence a bluffer can hardly be said to be practising a deception.

56 Which perhaps serves to illustrate the dangers in using an axiomatically based argument, as in part I, section 5.4, without careful attention to the interpretation of the axioms.

57 As always, the idea of active randomization can be replaced by uncertainty in the minds of other players as to what action is planned (see section 6.4).

58 Think of the flow of objective data as fueling an evolutive libration which progressively diminishes the capacity of rational players to "agree to disagree."

59 For example, if two different persons occupied the same player role in two separate and distinct occurrences of the same game.

60 Although, of course, there must be *some* uncertainty about the manner in which data are processed because of the Gödelian problems discussed in part I.

REFERENCES

Abreu, D., and Rubinstein, A. 1986: "The structure of Nash equilibrium in repeated games with finite automata," ICERD Discussion Paper 86/141, London School of Economics.

Aumann, R. 1974: "Subjectivity and correlation in randomized mixed strategies," *Journal of Mathematical Economics* 1, 67–96.

Aumann, R. 1981: "Survey of repeated games," in *Essays in Game Theory and Mathematical Economics in Honor of Oskar Morgenstern.* Mannheim/Wien/Zurich: Wissenschaftsverlung Bibliographisches Institut.

Aumann, R. 1987: "Correlated equilibrium as an expression of Bayesian rationality," *Econometrica* 55, 1–18.

Aumann, R., Katznelson, Y., Radner, R., and Rosenthal, R. 1981: "Approximate purification of mixed strategies," Bell Laboratories Economics Discussion Paper 192.

Axelrod, R. 1984: *The Evolution of Cooperation.* New York: Basic Books.

Banks, J., and Sobel, J. 1985: "Equilibrium selection in signalling games," mimeo, Massachusetts Institute of Technology, Mass.

Binmore, K. G. 1975: "An example in group preference," *Journal of Economic Theory* 10, 377–85.

Binmore, K. G. 1983: *Calculus.* Cambridge: Cambridge University Press.

Binmore, K. G. 1986: "Modeling rational players," ST/ICERD Discussion Paper 86/133, London School of Economics.

Binmore, K. G., and Dasgupta, P. 1987: *The Economics of Bargaining.* Oxford: Basil Blackwell.

Dawkins, R. 1976: *The Selfish Gene.* Oxford: Oxford University Press.

Harsanyi, J. 1967–8: "Games of incomplete information played by Bayesian players," parts I, II, and III, *Management Science* 14, 159–82, 320–34, 486–502.

Harsanyi, J. 1973: "Games with randomly disturbed payoffs: a new rationale for mixed strategy equilibrium points," *International Journal of Game Theory* 2, 1–23.

Harsanyi, J. 1975: "The tracing procedure," *International Journal of Game Theory* 5, 61–94.

Harsanyi, J., and Selten, R. 1980: "A non-cooperative solution concept with cooperative applications," chapter 1, draft, Center for Research in Management, Berkeley, Calif.

Harsanyi, J., and Selten, R. 1982: "A general theory of equilibrium selection in games," chapter 3, draft, Bielefeld Working Paper 1114, Bielefeld.

Hofstadter, D. 1983: "Metamagical themes," *Scientific American* 248, 14–28; 249, 14–20.

Hopcraft, J., and Ullman, R. 1979: *Introduction to Automata Theory, Languages, and Computation.* Reading, Mass.: Addison-Wesley.

Howard, N. 1971: *Paradoxes of Rationality: Theory of Metagames and Political Behavior.* Cambridge, Mass.: Massachusetts Institute of Technology.

Kohlberg, E., and Mertens, J. 1986: "On the strategic stability of equilibria," *Econometrica* 54, 1003–37.

Kreps, D. 1985: "Signalling games and stable equilibria," mimeo, Stanford University, Calif.

Kreps, D., and Wilson, R. 1982a: "Sequential equilibria," *Econometrica* 50, 863–94.

Kreps, D., and Wilson, R. 1982b: "Reputations and imperfect information," *Journal of Economic Theory* 27, 253–79.

Lewis, D. 1976: *Counterfactuals.* Oxford: Basil Blackwell.

Maynard Smith, J. 1982: *Evolution and the Theory of Games.* Cambridge: Cambridge University Press.

Marschak, T., and Selten, R. 1978: "Restabilizing responses, inertia supergames and oligopolistic equilibria," *Quarterly Journal of Economics* 92, 71–93.

Megiddo, N. 1986: "Remarks on bounded rationality," IBM Research Paper RJ 5270.

Megiddo, N., and Wigderson, A. 1986: "On play by means of computing machines," in *Theoretical Aspects of Reasoning about Knowledge*, ed. J. Halpern, pp. 259–74. Los Altos, Calif.: Morgan Kaufman.

Monod, J. 1972: *Change and Necessity.* Glasgow: Collins.

Myerson, R. 1984: "Sequential correlated equilibria of multistage games," J. Kellogg Graduate School of Management Discussion Paper 590, Northwestern University, Ill.

Nash, J. 1951: "Non-cooperative games," *Annals of Mathematics* 54, 286–95.

Neyman, A. 1985: "Bounded complexity justifies cooperation in the finitely repeated prisoner's dilemma," *Economic Letters* 19, 227–9.

Putnam, H. 1975: *Mind, Language, and Reality.* Cambridge: Camgbridge University Press.

Radner, R. 1980: "Collusive behavior in non-cooperative epsilon equilibria of oligopolies with long but finite lives," *Journal of Economic Theory* 22, 136–54.

Rubinstein, A. 1985: "Finite automata play the repeated prisoner's dilemma," ST/ICERD Discussion Paper 85/109, London School of Economics.

Scriven, M. 1965: "An essential undecidability in human behavior," in *Scientific Psychology: Principles and Approaches*, ed. B. Wolman. New York: Basic Books.

Selten, R. 1975: "Re-examination of the perfectness concept for equilibrium in extensive games," *International Journal of Game Theory* 4, 22–5.

Selten, R. 1978: "Chain-store paradox." *Theory and Decision* 9, 127–59.

Selten, R., and Leopold, U. 1982: "Subjunctive conditionals in decision theory and game theory," in *Philosophy of Economics*, vol. 2, *Studies in Economics*, ed. Stegmuller, Balzer, and Spohn. Berlin: Springer-Verlag.

Simon, H. 1955: "A behavioral model of rational choice," *Quarterly Journal of Economics* 69, 99–118.

Simon, H. 1959: "Theories of decision-making in economics," *American Economic Review* 49, 253–83.

Simon, H., 1976: "From substantive to procedural rationality," in *Method and Appraisal in Economics*, ed. S. Latis, pp. 129–48. Cambridge: Cambridge University Press.

Simon, H., 1977: *Models of Discovery*. Dordrecht: Reidel.

Von Neumann, J., and Morgenstern, O. 1944: *Theory of Games and Economic Behavior*. Princeton: Princeton University Press.

Index

Note: Page references in *italics* indicate figures.

Index prepared by Meg Davies (Society of Indexers)